Functional Phosphate Materials and Their Applications

Edited by Sadia Ameen, Mohammad Shaheer Akhtar and Hyung-Shik Shin

Published in London, United Kingdom

Functional Phosphate Materials and Their Applications
http://dx.doi.org/10.5772/intechopen.102267
Edited by Sadia Ameen, Mohammad Shaheer Akhtar and Hyung-Shik Shin

Contributors
Arturo Solís Herrera, María del Carmen Arias Esparza, Martha Patricia Solís Arias, Javier Fernandez, Irene García Cano, Raisha García-Cadme, Oscar Castaño, Tejas J., Siva Kumar V., Shruthi K. Siva Kumar, Ipsita Debat, Viswanathan K. Gowda, Nouha Mezned, Alok Singh Jayara, Rajeew Kumar, Priyanka Pandey, Manoj Kumar Bhatt, Sharad Pandey, Roshan Lal Meena, Arkady Mustaev, Sadia Ameen, Mohammad Shaheer Akhtar

First published in London, United Kingdom, 2023 by IntechOpen
IntechOpen is the global imprint of INTECHOPEN LIMITED, registered in England and Wales, registration number: 11086078, 5 Princes Gate Court, London, SW7 2QJ, United Kingdom

British Library Cataloguing-in-Publication Data
A catalogue record for this book is available from the British Library

Additional hard and PDF copies can be obtained from orders@intechopen.com

Functional Phosphate Materials and Their Applications
Edited by Sadia Ameen, Mohammad Shaheer Akhtar and Hyung-Shik Shin
p. cm.
Print ISBN 978-1-80356-800-3
Online ISBN 978-1-80356-801-0
eBook (PDF) ISBN 978-1-80356-802-7

Meet the editors

Professor Sadia Ameen obtained a Ph.D. in Chemistry from Jamia Millia Islamia, New Delhi, India. She is an associate professor in the Department of Bio-Convergence Science, Jeonbuk National University, Republic of Korea. Her current research focuses on dye-sensitized solar cells, perovskite solar cells, organic solar cells, sensors, catalysts, and optoelectronic devices. She specializes in the synthesis and application of clean energy materials. She received a gold medal for academic achievement and several scientific honors, including the Excellence in Research Award, Outstanding Scientist Award, Asia's Top-50 Scientist Award, Certificate of Excellence, and Best Researcher Award. She was listed among the world's top 2% of scientists by Stanford University, USA. She has authored or co-authored more than 130 peer-reviewed papers on solar cells, catalysts, and sensors, as well as contributed to book chapters and edited several books.

Professor M. Shaheer Akhtar obtained his Ph.D. in Chemical Engineering from Jeonbuk National University, Republic of Korea, where he is a full professor. His research interests include the photo-electrochemical characterizations of thin film semiconductor nanomaterials, composite materials, polymer-based solid-state films, solid polymer electrolytes and electrode materials for dye-sensitized solar cells (DSSCs), hybrid organic–inorganic solar cells, small molecule-based organic solar cells, and photocatalytic reactions.

Professor Hyung-Shik Shin received a Ph.D. in Chemical Engineering from Cornell University, USA, in 1984. He is an Emeritus Professor in the School of Chemical Engineering, Jeonbuk National University, Republic Korea, and president of the Korea Basic Science Institute (KBSI). He has been a visiting professor and invited speaker at several universities worldwide. He is an active executive member of various renowned scientific committees such as the Korean Institute of Chemical Engineers (KIChE), copyright protection, Korea Atomic Energy Research Institute (KAERI), and others. He has extensive experience in electrochemistry, renewable energy sources, solar cells, organic solar cells, charge transport properties of organic semiconductors, inorganic–organic solar cells, biosensors, chemical sensors, nano-patterning of thin film materials, and photocatalytic degradation.

Contents

Preface

Phosphate minerals are essential to the production of food and trigger algal blooms in lakes, rivers, and oceans. It is crucial to investigate phosphate minerals from economic, agricultural, environmental, and health perspectives. The sustainability of our well-being on the Earth depends on our ability to comprehend this key mineral and how we have utilized it over time.

This book provides a comprehensive overview of phosphate minerals, making it a handy reference for researchers looking for reliable fundamental knowledge. Geologists, chemists, environmental scientists, and engineers who are interested in the complex world of phosphorus will find this book to be of interest. It is also a useful book for those without specialized knowledge of the field because it is thorough and easy to read. It covers topics such as the surface functionalization of hydroxyapatite to prevent bacterial colonization, the intrinsic property of phosphorous to dissociate the water molecule, alternatives to soluble phosphorus fertilizers, and the use of plasma pseudocholinesterase as a predictor of mortality in organ transplant patients. Chapters present cutting-edge concepts and research trends and include helpful figures and tables.

We would like to express our gratitude to all the authors, rights holders, and reviewers for their contributions and feedback. We would not have been able to publish this book without their help. We would also like to thank IntechOpen for their support and encouragement during the preparation and publication of this book.

Sadia Ameen
Associate Professor,
Advanced Materials and Devices Laboratory,
Department of Bio-Convergence Science,
Advanced Science Campus,
Jeonbuk National University,
Jeonju, Republic of Korea

Mohammad Shaheer Akhtar
Professor,
New and Renewable Energy Material Development Center (NewREC),
Jeonbuk National University,
Jeonju, Republic of Korea

Hyung-Shik Shin
President,
Korea Basic Science Institute (KBSI),
Gwahak-ro, Yuseong-gu, Daejon, Republic of Korea

Chapter 1

Introductory Chapter: An Overview of Phosphate Mineral and Electrochemical Detection of Phosphate for Environmental Remediation

Sadia Ameen and Mohammad Shaheer Akhtar

1. Introduction

A class of inorganic salts, derived from sustainable phosphoric acid, is known as phosphate minerals. Over 200 different phosphate mineral classes have been identified to date, and all of them feature isolated (PO_4) tetrahedral units in their structure. Tetrahedrally coordinated phosphate (PO_4^{3-}), along with occasionally substituting arsenate (AsO_4^{3-}) and vanadate (VO_4^{3-}), chloride (Cl^-), fluoride (F^-), and hydroxide (OH^-) that also fit into the crystal structure, is present in phosphate minerals. Although the phosphate class of minerals is a sizable and varied group, only a few species are comparatively widespread. Phosphates can be divided into three categories: (i) primary phosphates, which have formed from a liquid; (ii) secondary phosphates, which have emerged as a result of the repeated occurrence of primary phosphates; and (iii) fine-grained rock phosphates. These phosphates have mostly developed from the sea containing phosphorus-bearing organic material at low temperatures.

Approximately 15–20% of the world's phosphate resources are thought to come from volcanic and weathered deposits, with the remaining 75% originating from sedimentary, marine rock formations. Aqueous fluids produced during the latter phases of crystallization are often where primary phosphates form. The granitic pegmatites are the common examples of the primary phosphates such as apatite [$Ca_5(F,Cl,OH)(PO_4)_3$], triphylite [$LiFePO_4$], lithiophilite [$LiMnPO_4$], and the rare-earth phosphates monazite [$(LaCe)(PO_4)$] and xenotime [$Y(PO_4)$]. Carbonatites and nepheline syenites are examples of ultramafic rocks, which are very low in silica and frequently include primary phosphates. Both impure limestones and calc-silicate rocks contain metamorphic apatite. The formation of secondary phosphates in different oxidation states can occur in water at low temperatures. Iron and manganese are typically present in both their divalent and trivalent oxidation forms, which results in vibrant hues. The phosphates such as strengite [$Fe(PO_4)(H_2O)_2$] and vivianite [$Fe_3(PO_4)_2(H_2O)_8$] are two typical species. There are various varieties of phosphate minerals, as follows:

IntechOpen

1.1 Pseudomorph mineral

Pseudomorph minerals are created when another substance undergoes chemical or structural change while preserving its original outward shape. The majority of pseudomorphs are granular and waxy on the inside, lack a regular cleavage, and appear to be crystalline, but they really exhibit optical properties that are distinct from those needed for their outer appearance. Pseudomorphs can be produced by putting the crystals of one mineral on top of the crystals of another. Alteration pseudomorphs can be created in a variety of ways, (i) through a modification in internal structure of the crystal without any modification in chemical composition (these pseudomorphs are known as paramorphs, e.g., aragonite changes to calcite and brookite changes to rutile), (ii) *via* the removal of a component from the original mixture (e.g., cuprite loses oxygen to form copper), and (iii) through the introduction of an ingredient to the (e.g., feldspar loses potassium silicate and gains water to become kaolinite).

1.2 Triplite mineral

It is a phosphate mineral comprising of Mn, Fe, Mg, and Ca phosphate [(Mn, Fe, Mg, Ca)$_2$PO$_4$(F,OH)], named as Triplite mineral. This mineral normally occurs in several parts of globe, for example, Bavaria, Ger.; Kimito, Fin.; Karibib, Namibia; and Maine, Connecticut, and Colorado in the United States, and notably, it is present in granite pegmatites as brightly colored (brown, salmon, flesh-red) masses.

1.3 Fluorapatite mineral

The fluorapatite mineral, also known as $Ca_5(PO_4)_3F$, is a common phosphate mineral. It can be found in many igneous rocks as tiny, frequently green, glassy crystals as well as magnetite deposits, hot hydrothermal veins, and metamorphic rocks. Additionally, the collophane is found in marine deposits.

1.4 Borate mineral

Borate mineral is a naturally occurring boron and oxygen combination. Borate minerals are generally rare; however, some can be found in significant deposits that can be mined for profit. The BO_3 triangle or BO_4 tetrahedron wherein oxygen or hydroxyl species are located at the triangle vertices or at the tetrahedron corners with a central boron atom, respectively, is incorporated into the structures of borate minerals. There may be both kinds of units in a single construction. Extended boron-oxygen networks can be formed by vertices sharing an oxygen atom, or they can contain a hydroxyl group if they are bound to another metal ion. Any given mineral's boron-oxygen complex shrinks in size as the temperature and pressure at which it forms rises and falls, respectively.

1.5 Tributyl phosphate

Tributyl phosphate is an organic liquid solvent used as a heat-exchange medium, a solvent for nitrocellulose, and the extraction of uranium and plutonium salts from reactor effluents. A phosphorus-containing substance with the chemical formula $(C_4H_9)_3PO_4$ is created when butyl alcohol and phosphorus oxychloride combine. Tributyl phosphate irritates the mucous membranes and corrodes the skin.

1.6 Amblygonite mineral

Amblygonite comprising of Li, Na, and Al phosphate [(Li,Na)$AlPO_4$(F,OH)] is phosphate mineral, which is extracted from ore of Li. It is often obtained from phosphate of lithium, that is, phosphate-rich granitic pegmatites having a very large crystal, white in color, and translucent masses. It has been mined at Keystone, South Dakota, as well as in a number of other nations, such as Zimbabwe and South Africa.

1.7 Cellophane mineral

Massive cryptocrystalline apatite, often fluorapatite or fluorian hydroxylapatite, makes up the majority of the fossil bone and phosphate rock known as collophane. It is typical to find horn-shaped concretions that are grayish-white, yellowish, or brown in hue.

1.8 Vanadate mineral

Vanadate is a naturally occurring mineral composed of vanadium (V), oxygen (O), and other metals. The majority of mentioned minerals are unusual and crystallized under highly specific circumstances, making them rare. Even though carnotite and vanadinite are occasionally mined as uranium and vanadium ore, respectively, most vanadates are of minimal economic significance; yet, mineral collectors esteem them for their vivid hues.

1.9 Sulfide mineral

Any member of the sulfur family-based compounds with one or more metals is referred to as a sulfide mineral. The majority of sulfides have straightforward structural characteristics, great crystallographic symmetry, and numerous metal-like characteristics, such as cluster of metals and electrical conductivity. They usually have high specific gravities, vivid hues, and low hardness. The general chemical formula A_mS_n, where A represents a metal, S ascribes to sulfur, can be used to indicate the composition of sulfide minerals. This formula yields the stoichiometries A_2S, AS, A_3S_4, and AS_2. Fe, Cu, Ni, Pb, Co, Ag, and Zn are the metals that are most frequently found in sulfides, while roughly 15 other metals can also enter sulfide structures.

1.10 Electrochemical detection of phosphate

The management of phosphorus nutrients is currently seen as a highly important societal task with significant implications for the economy, the environment, health, and industry, as a part of phospholipids, nucleic acids, or adenosine triphosphate, which are connected to cell membranes, genetic information storage and retrieval, and energy sources for cells, respectively. Phosphorus is in fact a crucial chemical element in live cells. Inorganic phosphate is produced in large quantities (82%) for use as fertilizers in agricultural fields, where the majority is lost to the environment [1, 2].

To solve environmental, financial, and health issues linked to phosphate processing, it is evident that there is a significant demand for quick, dependable, and affordable detection systems for continuous measurement. A variety of analytical techniques, including ion chromatography [3, 4], luminescence/fluorescence sensing [5–7], biosensing [8], and electro-analytical techniques [9–11], have been developed.

Here, we give a brief summary of recent advancements in the design of nanomaterials to meet the needs of selectivity and sensitivity for potentiometric and amperometric sensors, or biosensors, for phosphate measurement in actual waters.

2. Metal-based electrodes

Xiao et al. [12] introduced the unique cobalt-linked electrode for phosphate sensor. Due to particular interactions with the thin CoO layer generated at the electrode surface, solid-state Co-electrodes demonstrated a potentiometric response to H_2PO_4. This specific reactivity of the cobalt oxide surface and phosphate anions was recently validated by Ogata et al. [13]. Cobalt wires with a diameter of 1 mm were recently used to optimize a Co-based microsensor that can be used in lake water and soil samples with a few millimeters of spatial resolution [14].

3. Polymer-based sensors

In response to the electrochemical detection, despite the difficulties brought on by the hydrophilic nature of phosphate, ion selective membranes have been employed for phosphate detection [15]. The polyaniline film was doped with 0.5 M phosphonic acid and electrodeposited on a gold electrode [16]. According to Satoh et al. [17], an ionophore-doped polyvinyl chloride (PVC) membrane based on bis (dibromophenylstannyl) methane responds primarily to HPO_4^{2-} among different PO_4 species. The limitations of this sensor are the interference with OH^- and its short life-time (< 5 h). A new PVC membrane recently developed by Topcu et al. [18] was doped with a chitosan-clay combination. The as-prepared electrode after conditioning in Cr (III) solution expresses an anionic response, being particularly sensitive and selective toward HPO_4^{2-}.

4. Metal complex-based sensors

Applying some metal complexes including copper phthalocyanine (CuPc) [19–21] or M-2,6-bis(bis(2-pyridylmethyl)amino methyl)-4-methylphenol (M-BPMP, M = Zn and Cu) [22], uranyl salophene III [23] have already being used for the detection of phosphates.

Author details

Sadia Ameen[1*] and Mohammad Shaheer Akhtar[2]

1 Advanced Materials and Devices Laboratory, Department of Bio-Convergence Science, Jeongeup Campus, Jeonbuk National University, Republic of Korea

2 New and Renewable Energy Material Development Center (NewREC), Jeonbuk National University, Jeonbuk, Republic of Korea

*Address all correspondence to: sadiaameen@jbnu.ac.kr

References

[1] Schröder JJ, Cordell D, Smit AL, Rosemarin A. Sustainable Use of Phosphorus: Plant Research International. Netherland: Wageningen University and Research Centre - Stockholm Environment Institute (SEI); 2009

[2] Duffy G, Regan F. Recent developments in sensing methods for eutrophying nutrients with a focus on automation for environmental applications. Analyst. 2017;**142**:4355-4372

[3] Yokoyama Y, Danno T, Haginoya M, Yaso Y, Sato H. Simultaneous determination of silicate and phosphate in environmental waters using pre-column derivatization ion-pair liquid chromatography. Talanta. 2009;**79**:308-313

[4] Rantakokko P, Mustonen S, Yritys M, Vartiainen T. Ion chromatographic method for the determination of selected inorganic anions and organic acids from raw and drinking waters using suppressor current switching to reduce the background noise. Journal of Liquid Chromatography and Related Technologies. 2004;**27**:829-842

[5] Borse V, Jain P, Sadawana M, Srivastava R. Turn-on' fluorescence assay for inorganic phosphate sensing. Sensors and Actuators: B. 2016;**225**:340-347

[6] Mahajan PG, Desai NK, Dalavi DK, Bhopate DP, Kolekar GB, Patil SR. Cetyltrimethylammonium bromide capped 9-anthraldehyde nanoparticles for selective recognition of phosphate anion in aqueous solution based on fluorescence quenching and application for analysis of chloroquine. Journal of Fluorescence. 2015;**25**:31-38

[7] Kaur N, Kaur S, Kaur A, Saluja P, Sharma H, Saini A, et al. Nanoparticle-based, organic receptor coupled fluorescent chemosensors for the determination of phosphate. Journal of Luminescence. 2014;**145**:175-179

[8] Ameen S, Akhtar MS, Shin HS. Nanocages-augmented aligned polyaniline nanowires as unique platform for electrochemical non-enzymatic glucose biosensor. Applied Catalysis A: General. 2016;**517**:21-29

[9] Kim EB, Imran M, Lee EH, Akhtar MS, Ameen S. Multiple ions detection by field-effect transistor sensors based on ZnO@GO and ZnO@ rGO nanomaterials: Application to trace detection of Cr (III) and Cu (II). Chemosphere. 2022;**286**:131695

[10] Jang GS, Kim EB, Akhtar MS, Shin HS, Ameen S. An exploration of 3-methoxypropionitrile chemical sensor based on layered hexagonal $NiCo_2O_4$ nanoplates as electrode material. Ceramics International. 2021;**47**:15357-15366

[11] Ameen S. Colloidal synthesis of $NiMn_2O_4$ nanodisks decorated reduced graphene oxide for electrochemical applications. Microchemical Journal. 2021;**163**:105912

[12] Xiao D, Yuan HY, Li J, Yu RQ. Surface-modified cobalt-based sensor as a phosphate-sensitive electrode. Analytical Chemistry. 1995;**67**:288-291

[13] Ogata F, Imai D, Toda M, Otani M, Kawasaki N. Adsorption of phosphate ion in aqueous solutions by calcined cobalt hydroxide at different temperatures. Journal of Environmental Chemical Engineering. 2015;**3**:1570-1577

[14] Ding X, Behbahani M, Gruden C, Seo Y. Characterization and evaluation

of phosphate microsensors to monitor internal phosphorus loading in Lake Erie sediments. Journal of Environmental Management. 2015;**160**:193-200

[15] Warwick C, Guerreiro A, Soares A. Sensing and analysis of soluble phosphates in environmental samples: A review. Biosensors & Bioelectronics. 2013;**41**:1-11

[16] Huang Y, Ye Y, Zhao G, XiaominWu KY, Mur L, Jiwan Han Qin H. An all-solid-state phosphate electrode with H_3PO_4 doped polyaniline as the sensitive layer. International Journal of Electrochemical Science. 2017;**12**:4677-4691

[17] Satoh H, Miyazaki Y, Taniuchi S, Oshiki M, Rathnayake RMLD, Takahashi M, et al. Improvement of a phosphate ion-selective microsensor using Bis(dibromophenylstannyl) methane as a carrier. Analytical Sciences. 2017;**33**:825-830

[18] Topcu C, Caglar B, Onder A, Coldur F, Caglar S, Guner EK, et al. Structural characterization of chitosan-smectite nanocomposite and its application in the development of a novel potentiometric monohydrogen phosphate-selective sensor. Materials Research Bulletin. 2018;**98**:288-299

[19] Abbas MN, Radwan ALA, Nooredeen NM, Abd El-Ghaffar MA. Selective phosphate sensing using copper monoamino-phthalocyanine functionalized acrylate polymer-based solid-state electrode for FIA of environmental waters. Journal of Solid State Electrochemistry. 2016;**20**:599-1612

[20] Barhoumi L, Baraket A, Nooredeen NM, Ben AM, Abbas MN, Bausells J, et al. Silicon nitride capacitive chemical sensor for phosphate ion detection based on copper phthalocyanine - acrylate-polymer. Electroanalysis. 2017;**29**:1586-1595

[21] Zina F, Nooredeen NM, Azzouzi S, Ben AM, Abbas MN, Errachid A. Novel sensitive impedimetric microsensor for phosphate detection based on a novel copper phthalocyanine derivative. Analytical Letters. 2018;**51**:371-386

[22] Li L, Shang G, Qin W. Potentiometric sensing of aqueous phosphate by competition assays using ion-exchanger doped-polymeric membrane electrodes as transducers. Analyst. 2016;**141**:4573-4577

[23] Pankratova N, Ghahraman AM, Yuan D, Crespo GA, Bakker E. Local acidification of membrane surfaces for potentiometric sensing of anions in environmental samples. ACS Sensors. 2016;**1**:48-54

Chapter 2

Inorganic Phosphate: The Backbone of Life

Arkady Mustaev

Abstract

Inorganic phosphate (P_i) plays a crucial role in many biochemical pathways. Broad P_i involvement in the structure and function of biological entities reflects a striking unity of inorganic and organic matter in life processes. P_i functions as a constituent of cellular metabolites as well as a building material for bones in vertebrates and shells in some marine species, owing to the ability of P_i to form robust minerals. Dysregulation of phosphate metabolism causes serious medical disorders, such as osteoporosis, arterial medial calcification, hypophosphatemia, and kidney stone formation. The purpose of this chapter is to provide a brief but comprehensive overview of inorganic phosphate biology. The chapter aims at a broad audience that includes advanced graduate students and first-year medical students as well as researchers and scientists interested in the basics of P_i bioorganic, biophysical, bioinorganic, and biomedical chemistry. Herein, the author i) describe major P_i functions in current life forms; ii) highlight unique P_i properties that underpin its role in life processes, iii) rationalize the natural choice of P_i for design of biological molecules, and iv) discuss the possible involvement of inorganic phosphate and its minerals in events that led to the emergence of life.

Keywords: inorganic phosphate, cellular functions, biological phosphate minerals, bone metabolism, pathological calcification, catalysis, nanobacteria, life origin

1. Introduction

Inorganic phosphate, P_i, is the simplest and only form of phosphorus existing in nature (**Figure 1**); it plays a central role in cellular energetics and metabolism as well as in biological structure and regulation (for reviews see Refs. [1–3]). It is believed that P_i participated in the principal events that led to the origin of life [4–6]. Indeed, phosphate-containing compounds are essential constituents of all living cells.

Previous reviews covering particular aspects of P_i involvement in the life process (cited in the text) are available. This chapter fulfills the need for a broader outlook on the subject. This review compiles the most significant and recent data and presents their critical analysis from the perspective of a researcher who has an extensive experience in chemistry and biochemistry of inorganic phosphate and natural phospho-organic compounds.

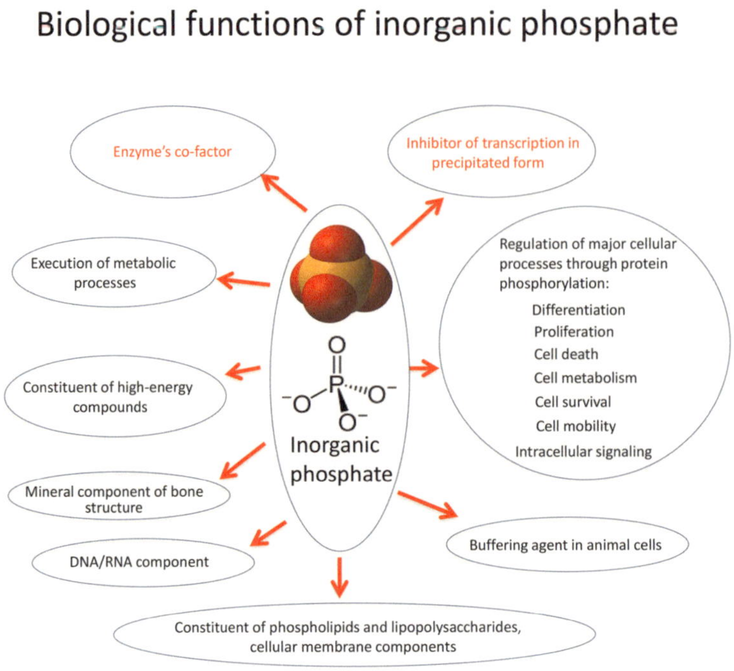

Figure 1.
Major functions of inorganic phosphate. Recently discovered functions are indicated by the red font.

Special emphases will be made on the properties of phosphate that define its function in major cellular processes as well as on the biologically important P_i minerals and biomineralization processes in vertebrate species. The later include bone metabolism as well as pathological P_i minerals deposition cases such as kidney stone formation and arterial medial calcification. A significant part of the review will be devoted to the exotic case of P_i biomineralization, which is the phenomenon of "nanobacteria". The importance of this phenomenon in regard to life origin and early evolution is not yet widely appreciated. Therefore, the author will discuss this topic in detail. Finally, the author will describe specific properties of P_i that likely determined the natural choice of this compound as part of life material.

2. Properties of phosphate that define its function in major cellular processes

Broad P_i involvement in life processes is explained by the unique properties of this element. For example, phosphorus is able to covalently associate with five other atoms by contributing outer shell electrons to the bond formation (e.g., PCl_5 structure). In the $(PO_4)^{3-}$ moiety, phosphorus forms single bonds with three oxygens and a double bond with the remaining oxygen (**Figure 1**). In the cell, phosphate forms anhydrides, esters, phosphamides, and diesters. Major cellular structures involving phosphate as a building block are listed in **Table 1**. Below, the author briefly describe the role of the phosphate group in these structures and highlight the involvement of P_i in major life processes.

DOI: http://dx.doi.org/10.5772/intechopen.109117

Metabolite	Structure	Function
RNA/DNA	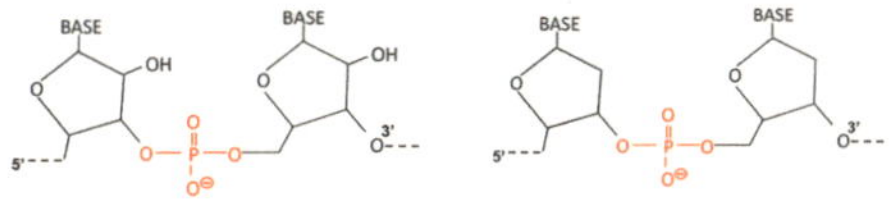	Genetic material
Adenosine triphosphate	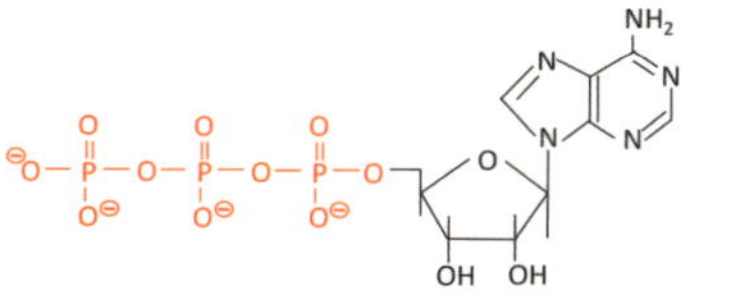	Cellular energy source, substrate for RNA synthesis
Creatine phosphate	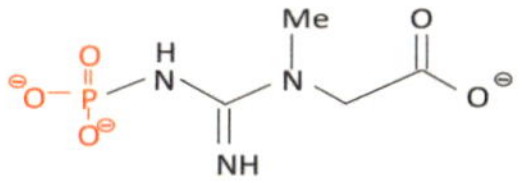	Emergency energy source, ATP production in anaerobic conditions
Polyphosphate	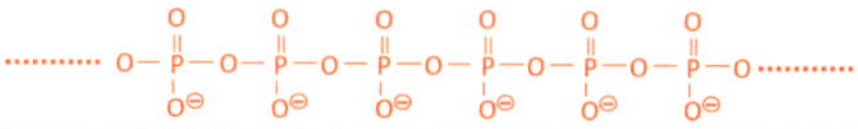	Phosphate storage, energy source
Aminoacyl-adenylate	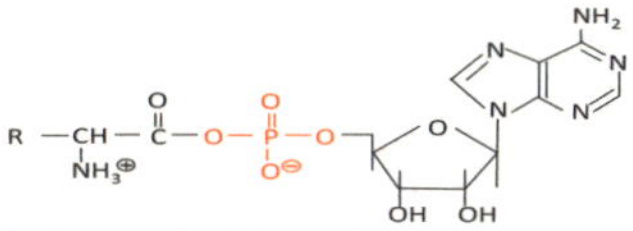	tRNA aminoacylation
1,3-diphosphoglycerate	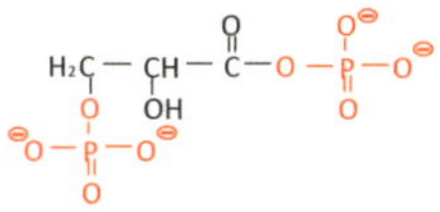	ATP production
NAD (NADP)	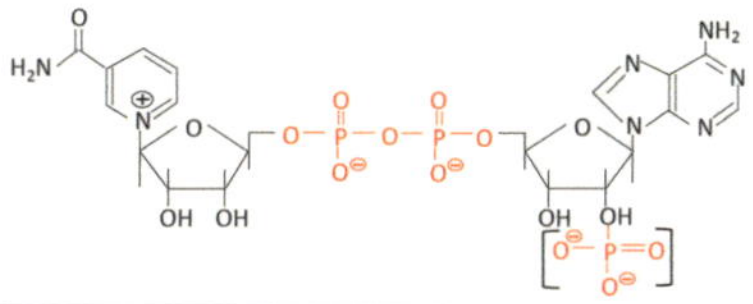	Redox reactions, ATP production, signaling, redox sensing, detoxification
Pyridoxal phosphate	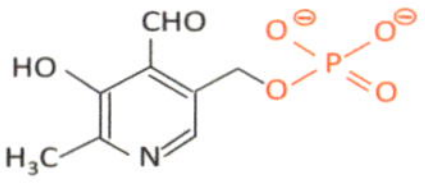	Coenzyme in the reactions of transamination, decarboxylation, deamination, and racemization of amino acids.
Inorganic phosphate	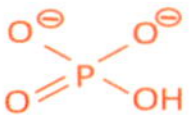	Coenzyme in transcript cleavage reaction by RNA polymerase. Substrate for ATP synthesis by F_1F_0 ATP synthase
cAMP	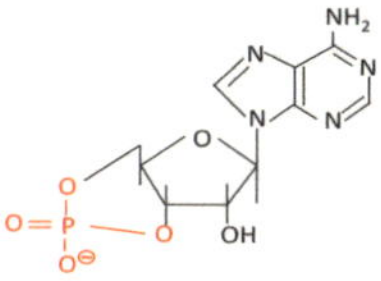	Intracellular signal transduction
Phospholipid	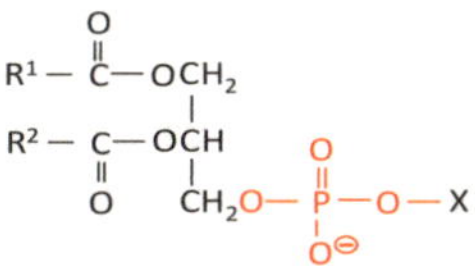	Component of cellular membranes

Metabolite	Structure	Function
Phosphoglycerate	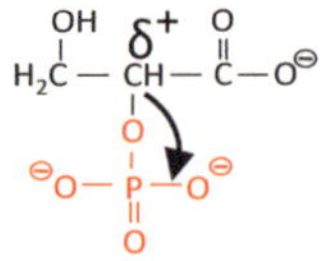	Metabolite activation

Table 1.
Major representatives of biological molecules that involve phosphate.

2.1 P_i role in structure of nucleic acids

Consideration of the major classes of phosphate-containing biological molecules (**Table 1**) suggests distinctive roles of phosphorus in their structure and function. For example, in nucleic acids, phosphate forms diester linkages that connect the nucleoside units. The linkages display high chemical resistance toward hydrolytic cleavage, providing the stability required for maintenance and storage of genetic information: The half-life for the uncatalyzed attack of water on the dimethyl phosphate monoanion (a model for DNA linkage) at ambient temperature is about 10^7 years [7]. RNA is about 10^4–10^5-fold less stable due to an intramolecular attack of the ribose 2′ hydroxyl group on the neighboring phosphodiester [8]. Nevertheless, RNA possesses remarkable resistance to hydrolysis [9]. The stability of nucleic acids is expected to have been a crucial factor at the beginning of life since the first life forms probably functioned at temperatures of 55–85°C [10], conditions that accelerate hydrolysis rate by a factor of 10^2–10^4 relative to ambient temperatures. In contemporary organisms, the integrity of genetic information is additionally ensured by enzymes that recognize and repair damaged DNA. Importantly, a stable negative charge at the phosphodiester linkage maintains the high solubility of nucleic acids. This is in contrast to the low solubility of nucleic acid analogs having a peptide backbone (PNA), which lacks a charge [11].

2.2 P_i anhydrides, mixed anhydrides, and phosphamides: Energy-rich compounds

A separate group of cellular phosphometabolites comprises reactive energy-rich compounds, such as phosphoanhydrides (nucleoside-5'di- and triphosphates, NDP, NTP), nicotinamide adenine di- or triphosphates, flavin adenine dinucleotide (NAD (H), NADP(H)), (FAD$(H)_2$), polyphosphate (polyP), mixed anhydrides (e.g., aminoacyl adenylates), as well as phosphamides, creatine, and arginine phosphates. Generally, the energy stored in these compounds is used to drive energetically disfavored biochemical processes in coupled reactions.

ATP is an example of targeted energy delivery in biological systems in which the energy stored in the inorganic triphosphate moiety is directed by the adenosine "address" to an enzyme executing a particular ATP-dependent metabolic process. ATP is a major energy source that serves multiple functions, including muscle contraction and biochemical synthesis (for review, see Ref. [12]). When consumed in metabolic reactions, it converts either to adenosine diphosphate (ADP) or to ade nosine monophosphate (AMP). While possessing high energy, phosphoanhydrides displays remarkable kinetic stability due to the significant activation barrier of the phosphoryl-transfer reaction. Thus, at 100°C the half-time time for non-catalyzed ATP hydrolysis is about 30 min. This relative stability allows the controlled release of the energy stored in ATP *via* enzyme-mediated processes in which enzymes direct the rate of energy release by lowering the activation barrier of the reaction. This relative stability of phosphoanhydride seems to have been of even greater significance during the origin of life, preserving the pool of these compounds at high temperatures. In addition, ATP (as well as other NTPs) is a substrate for RNA synthesis by RNA polymerase (RNAP), while the corresponding deoxy NTPs are utilized for DNA synthesis by DNA polymerases. The former reaction, executed by RNA enzymes (ribozymes), is believed to be central to the origin of life by supporting the self-duplication of these structures in the primordial "RNA world."

Inorganic polyphosphate (polyP), a polymeric P_i form found in all kingdoms of living organisms, is a phosphorus and energy source that ensures the survival of microorganisms under extreme conditions [13]. It also has an important regulatory role by inhibiting RNA polymerase-mediated transcription [14].

The NAD(P) class of compounds participates in numerous reactions involving electron transfer due to facilitated redox conversion of the nicotinamide moiety (for review, see Refs. [15, 16]). Even though the NAD family of metabolites possesses an energy-rich phosphoanhydride bond, it is rarely used to drive biochemical reactions. An example is NAD^+ involvement in the repair of DNA breaks in which the adenylyl moiety of the metabolite is transferred to the phosphate residue of a broken phosphodiester bond, thereby affording an active pyrophosphate derivative that phosphorylates the neighboring DNA 3′ hydroxyl residue to seal the break [17].

FAD is a redox-active cofactor, which assists various proteins (flavonoids), involved in enzymatic reactions. Flavoproteins employ the unique and versatile structure of flavin moieties to execute difficult redox reactions owing to the strong oxidizing potential of FAD, which is greater than that for NAD^+.

Creatine phosphate [18] possesses a phosphamide bond, whose energy (10.3 kcal/mol) exceeds even that for ATP hydrolysis (7.3 kcal/mol). It serves to quickly recharge ADP to ATP (through phosphoryl transfer) in the active muscles of vertebrate organisms. In invertebrates, this function is executed by the structurally analogous arginine phosphate.

Mixed anhydrides of phosphoric and carboxylic acids are another class of energy-rich compounds. They possess activated carbon and phosphorus atoms within an anhydride C-O-P segment; therefore, these compounds can display acyl or phosphoryl-transfer reactivity. The former reaction pathway in observed for aminoacyl-adenylates, which in an enzyme-mediated process transfers an *aminoacyl* residue to the 3′ hydroxyl group of tRNAs [19] for protein biosynthesis. In contrast, the mixed anhydride 1,3-diphosphoglycerate, a metabolite of glycolysis, transfers a *phosphoryl* group to ADP to produce ATP.

2.3 P_i functions exploiting its unique physicochemical properties

Some phosphate-bearing cellular compounds take advantage of negatively charged P_i residues to enhance metabolite recognition by proteins at their binding sites through salt bridging and hydrogen bonding. This appears to be the case for the NAD/NADP class of compounds, pyridoxal phosphate, and cyclic AMP. In addition, the incorporation of charged P_i moieties into the structure ensures metabolite retention by the cell *via* impeding diffusion through the cell membrane. Metabolite retention could have been an important factor in primordial organisms, whose cell membranes are thought to have been leakier than those of modern cells.

The inclusion of P_i residues in lipid components of cell membranes is due to the highly polar nature of the P_i group owing to the double-negative electric charge and therefore its strong water-hydration status. In addition, P_i residues as constituents of lipopolysaccharide (LPS) of outer membranes in gram-negative bacteria coordinate with Me^{2+} ions (e.g., Mg^{2+}), which stabilize outer membrane structures through the fusion of neighboring LPS units. In particular, this coordination limits antibiotic access to cells [20].

The attachment of phosphate residues to cellular proteins in the course of post-translational modification (PTM) is a significant biological process. It serves multiple purposes, including intracellular signaling, cell differentiation, cell death, survival, cell mobility, and regulation of cellular metabolism (for review, see [3]). Phosphorylation of proteins is one of the most observed PTMs. Phosphate esters of Ser, Thr, and Tyr are the most common modification products. The critical feature of these PTM entities—phosphoamino acids in proteins—is that they do not resemble any natural amino acid residue; therefore, they provide a special means of modulating the physicochemical properties of protein surfaces. In particular, the phosphate group, with its large, hydrated shell and dense negative charge, is physically quite distinct from the common negatively charged amino acids, Asp and Glu, whose carboxyl side chains possess only a single negative charge and as such, a smaller hydrated shell than phosphate. Consequently, a protein-linked phosphate group can form hydrogen bonds or salt bridges of greater stability with either Arg or Lys side chains [21]. Moreover, this moiety can bridge with Asp and Glu residues *via* coordination of the Mg^{2+} ion. As such, protein-linked phosphates can regulate function or the interaction with another protein or small molecule by i) inducing a conformational change within a protein, ii) promoting inducible protein-protein interactions through specific recognition of phosphoproteins by phosphospecific binding domains in other proteins, and iii) changing the subcellular location of a protein by affecting its transport. Moreover, tagging a protein with P_i can either prevent or promote protein degradation in a ubiquitin-dependent process [22].

A phosphate is an electronegative group [1, 2, 23], a property seen in various pathways to activate metabolites for biochemical reactions. This feature is exemplified by phosphoglycerate (**Table 1**), an intermediate in glycolysis. In this compound, electron withdrawal by the phosphate group induces a partial positive charge on the adjacent carbon atom, which activates the C-C bond in a subsequent dehydration step that produces phosphoenolpyruvate in a β-elimination reaction.

The ability of P_i and condensed phosphates (structures that include phosphoanhydride units) to coordinate divalent metal ions is a part of the catalytic mechanism underlying numerous biochemical reactions that are achieved through nucleotidyl or phosphoryl transfer. These reactions include nucleic acid biosynthesis, repair, and degradation. Coordination at the active center by Mg^{2+} ions (exemplified below by RNA polymerase) presents substrate groups for catalysis and promotes the above reactions (e.g., through electron withdrawal or stabilization of the reaction intermediates, or reaction transition states).

Another consequence of phosphate coordination is the poor solubility of P_i salts with the divalent metal ions (e.g., Ca^{2+} and Mg^{2+}). This property is exploited in the construction of bone and teeth but can also cause medical disorders (e.g., cardiovascular calcification and formation of kidney stones), which are described below.

2.4 P_i utilization as a substrate or a cofactor in biochemical reactions

P_i metabolism starts with its inclusion into cellular molecules in enzyme-mediated processes. P_i is used as a substrate in two principal reactions yielding energy-rich compounds. One is the direct phosphorylation of ADP, which produces ATP. This highly energy-consuming conversion, executed by ATP synthase, is driven by the electrochemical gradient created during respiration by the difference in proton (H^+) concentration across the inner mitochondrial membrane in eukaryotes and the plasma membrane in bacteria. The second reaction is NAD^+-mediated formation of 1,3-bisphosphoglycerate from P_i and glyceroaldehyde-3-phosphate in the course of glycolysis. Phosphoryl transfer from this compound to ADP affords ATP in the following step.

A function of inorganic phosphate recently discovered by our group is the ability of P_i to serve as an enzyme cofactor [24]. Since this property may reflect unknown P_i involvement in cell biology, the author consider the phenomenon in detail. RNA, synthesized by DNA-depended RNA polymerase (RNAP) from a DNA template as the first step in gene expression, encodes the genetic information that translates to the corresponding protein structure at the following step. Although RNA synthesis is the principal activity of RNAP, the enzyme can also degrade transcripts in a reaction that eliminates the 3′ terminal RNA nucleotide by hydrolysis [25]. We found that P_i strongly stimulates this RNAP reaction, suggesting a potential role for this anion in gene expression.

The stimulating P_i effect can be rationalized in the context of RNAP catalytic mechanism (**Figure 2**). Both RNA synthesis and degradation by RNAP occur at the same active center [26] and proceed through an S_n2 mechanism [25–28], which is common to all nucleic acid polymerases and ribozymes. It is thought to have supported RNA self-duplication in the primordial RNA world [29–32]. The mechanism involves two Mg^{2+} ions (**Figure 2**) chelated by an aspartate triad in the active center of the enzyme. One Mg^{2+} ion (Mg-I) is permanently retained within the enzyme, whereas the other (Mg-II) binds transiently; it must be recruited through additional coordination for each round of catalysis [25]. Thus, Mg-II is stabilized by chelation with the phosphate groups of the NTP substrate at the nucleotidyl transfer step of RNA synthesis (**Figure 2A**). In this process, Mg-I activates the 3′- hydroxyl of the terminal RNA residue (by promoting its ionization) and the α-phosphate of an NTP substrate for nucleophilic attack; Mg-II facilitates the release of pyrophosphate (**Figure 2A**) by electron withdrawal through coordination. The two Mg^{2+} ions switch

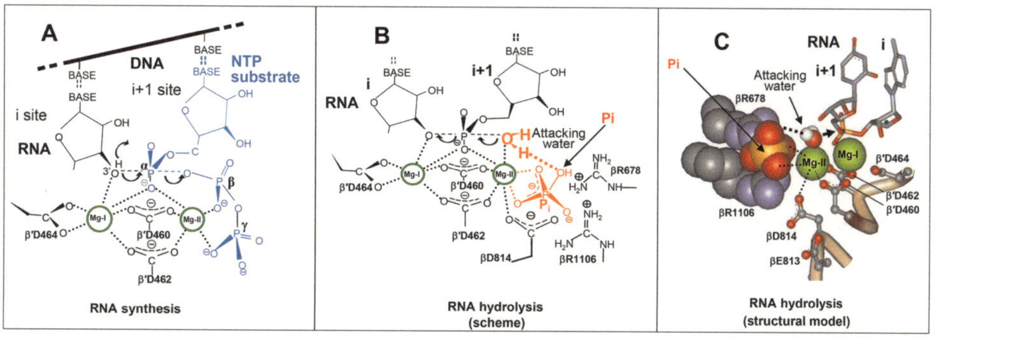

Figure 2.
Modes for the RNAP active center action. (A) Nucleotidyl-transfer reaction. (B) Proposed mechanism for P_i action in hydrolytic RNA cleavage. Amino acid residues of the active center involved in the reactions of A and B are indicated. (C) Structural model for B. A part of the figure is from [24].

roles in the hydrolytic cleavage of RNA (**Figure 2B**). Thus, Mg-II activates the attacking water, while Mg-I aids in the release of the leaving group. We found that inorganic phosphate accelerates hydrolytic RNA cleavage about 2000-fold by stabilizing Mg-II at the active center through additional coordination and by activating the attacking water molecule (**Figure 2B** and **C**). In this capacity, P_i is the smallest known prosthetic group assisting an enzymatic reaction.

The above finding represents a stunning example of how simple cellular molecules can reprogram an enzyme's active center by providing functional groups for catalysis. The main P_i features as cofactor in this reaction involve the P_i ability to i) salt-bridge with positively charged active center amino acid side chains, ii) coordinate catalytic Mg^{2+} ion, and iii) form a hydrogen bond. Strikingly, inorganic phosphate mimics, such as arsenate $(HAsO_4)^{2-}$ and vanadate $(HVO_4)^{2-}$ anions, stimulate this reaction to the same extent as P_i.

3. Role of inorganic phosphate in biomineralization

3.1 General concepts and factors involved in mineral formation in living organisms

Phosphate compounds play a central role in biologically controlled mineralization (biomineralization), the processes by which living organisms produce minerals, often to harden or stiffen existing tissues (for review see [33]). The formation of hard tissues represented an evolutionary breakthrough by providing enhanced body strength, advanced mobility, and protection. It is a widespread phenomenon observed in all six taxonomic kingdoms; it is about 500 million years old. In this section, the author concentrate on biomineralization occurring in humans and other vertebrate species. In these organisms, mineral deposition is used to form hard structures, such as bones and teeth. In the cases of pathology, biomineralization develops in soft tissues causing illnesses, such as formation of stones in kidneys and bone-like deposits in heart valves (e.g., arterial medial calcification). A significant number of these pathological cases involve the precipitation of poorly soluble phosphate forms.

Solubility, a fundamental property of compounds, is defined as the maximal concentration of the compound in solution. Two distinct kinds of solubility exist i) equilibrium (thermodynamic) solubility, a type of chemical equilibrium between a compound in the solid state and a solution of that compound, and ii) kinetic solubility, which occurs when the concentration of the compound in solution exceeds its equilibrium (thermodynamic) solubility, thus creating a supersaturated solution. Even though supersaturated solutions are unstable, some compounds do not precipitate over prolonged times, especially in media containing other compounds (e.g., intracellular environment or blood).

Crystallization affords a solid state in which atoms or molecules are highly organized into an assembled structure known as a crystal. This process involves two major events: *nucleation* and *crystal growth*, which are driven by thermodynamics as well as by the physicochemical properties of the compounds. Nucleation is the step in which the solute molecules or atoms, randomly dispersed in the solvent, start to gather into nano-scale clusters in a small region. These clusters are metastable owing to their ability to dissolve in an equilibrium process and their need to reach a critical size in order to become stable nuclei. At this stage, the atoms or molecules arranged in an organized and periodic manner that defines the crystal structure. Crystal growth is the subsequent size increase of nuclei that succeeds in achieving the critical cluster size.

Supersaturation is a prerequisite of crystallization by making the process thermodynamically favorable. Depending on conditions, either nucleation or growth of existing crystals may predominate, thereby controlling crystal size. Both steps of crystallization are highly sensitive to the presence of other molecules or objects in the medium, which can either promote or inhibit the process. This can be illustrated by the school chemistry demonstration in which a thread placed into a supersaturated solution turns overnight into a magnificent string of beads—crystals.

Crystallization occurring in living organisms has several important specifics that make it highly controllable. The major compound in biomineralization is calcium phosphate [34], which can crystallize in various forms (see **Table 2**), the most common being hydroxyapatite (HAP). HAP kinetic solubility allows its accumulation at levels sufficient to create supersaturated solutions, while the thermodynamic solubility of this compound enables controllable deposition in biological systems. Solubility is determined by the chemical interactions of the constituents both within the crystal unit and between the units. In the case of HAP, these interactions are represented by ionic bonds between the metal cation and negatively charged oxygens of the phosphate group and the hydroxyl group as well as by coordination bonds between Ca^{2+} and the above groups (**Figure 3**). Solubilization of the crystal destroys all the bonds, releasing the constituents of the mineral as hydrated free ions. The sustainable solubility of HAP can be attributed to the moderate strength of the coordination bonds between the Ca^{2+} ion and the interacting anionic groups of the mineral. Indeed, the solubility of phosphate minerals with other divalent metals that

Name, formula	Appearance	Solubility	Ref.
Magnesium phosphate $Mg_3(PO_4)_2$		10^{-5} M 0.5×10^{-5} M	[35] [36]
Magnesium hydrogen phosphate trihydrate. Newberyite $MgHPO_4 \cdot 3H_2O$		1.3×10^{-3} M 1.44×10^{-3} M	[37] [38]
Tricalcium phosphate (TCP) $Ca_3(PO_4)_2$		0.5×10^{-6} M	[36]
Calcium hydrogen phosphate. Monetite $CaHPO_4$		0.3×10^{-3} M	[39]

Name, formula	Appearance	Solubility	Ref.
Dicalcium phosphate dehydrate. Brushite $CaHPO_4{\cdot}2H_2O$		0.5×10^{-3} M	[39]
Hydroxylapatite (HAP) $Ca_5(PO_4)_2OH$		$0.8\text{–}1.6 \times 10^{-7}$ M	[40]
Struvite $Mg(NH_4)PO_4$	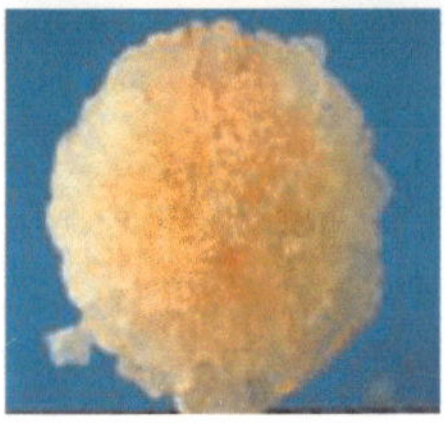	0.5×10^{-4} M	[41]

Table 2.
Solubility and appearance of biologically relevant phosphate compounds and minerals.

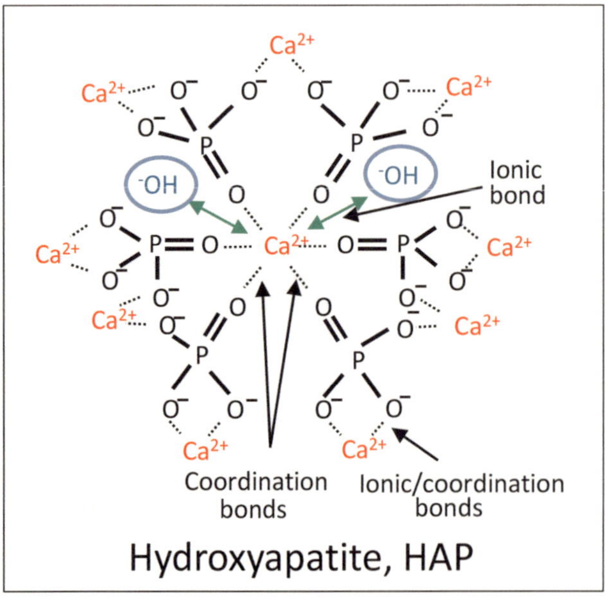

Figure 3.
Hydroxyapatite structure. Various types of bonds within the HAP structural unit $Ca_{10}(PO_4)_6(OH)_2$ are indicated.

form stronger coordination bonds is lower. This could be why (along with the high environmental availability of soluble Ca^{2+} forms) this metal evolved for biomineralization as a P_i counterpart. Notably, Mg^{2+} and carbonate ions (about 0.6% and 6%, respectively) incorporate into a HAP structure upon deposition.

The second important aspect of biomineralization is that it occurs in complex environments that include both biopolymers and small compounds that can interfere with HAP deposition. This enables control of biomineralization by the maintenance of the balance between metabolites that either inhibit or promote crystallization [42]. Importantly, the concentration of Ca^{2+} and P_i ions in biological tissues is well above the saturation level. Therefore, the deposition of these ions in the form of HAP is normally prevented by inhibitors. However, in pathological situations, the elevated concentration of the two ions can cause uncontrollable HAP deposition ([43] and below), which underscores the importance of maintaining the ions at the optimal level.

3.2 Bone formation

Bone is a rigid organ that constitutes the skeleton of vertebrate animals. Bones protect the body's organs, produce red and white blood cells, store minerals, and provide structure and support for the body, thereby enabling mobility. Structurally, bone represents a composite material comprised of a matrix of collagen fibrils surrounded by mineral deposit hydroxyapatite (**Figure 4**). The composite structure provides remarkable rigidity and resistance to mechanical stress. The relative strength of bone exceeds that of concrete. The human thighbone can sustain roughly 6000 lbs. of compressive force [45]. Bone metabolism is a highly dynamic process (for review see [46]) that involves bone formation as well as bone resorption, which is the solubilization of the bone material. Both processes are controlled by specialized cells populating bone tissues. Osteoblasts and osteocytes are involved in the formation and

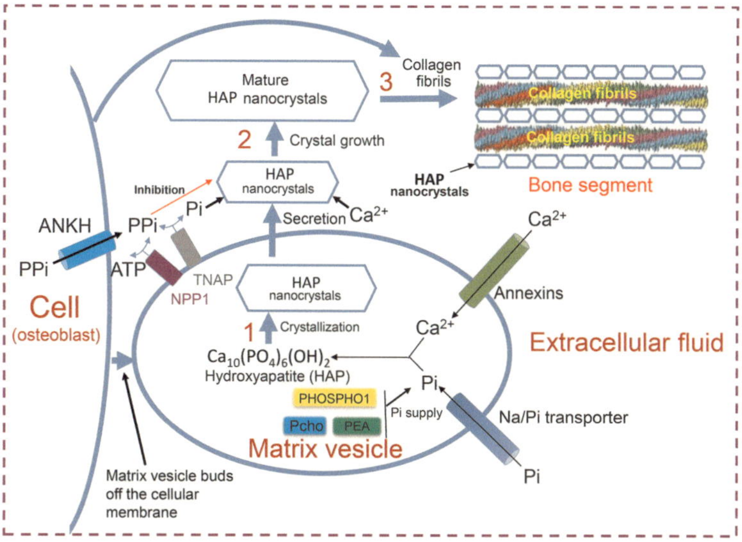

Figure 4.
The process of bone formation. This starts with budding the matrix vehicle (MV) of the osteoblast cell. The creation of high concentration of Ca^{2+} and pi inside MV initiates HAP formation (step 1) followed by the excretion of the nanocrystals through the membrane and their further growth in the extracellular fluid (step 2). Mature HAP crystals then associate with collagen fibrils to form mineralized bone material (step 3). This is an extended and modified version of the original figure of Ref. [44].

mineralization of bone, while osteoclasts execute the resorption of bone tissue. **Figure 4** shows the mineralization process mediated by osteoblasts [42, 44, 46]. As stated above, initiation of crystallization requires high concentrations of Ca^{2+} and P_i ions, which if generated intracellularly would interfere with vital cellular processes. Osteoblasts produce specialized matrix vesicles (MV) that are "factories," where hydroxyapatite nanocrystals are manufactured by creation of optimal conditions for crystallization. Matrix vesicles are small formations, 100–200 nm in diameter, equipped with a variety of features that are involved in generation of high concentrations of Ca^{2+} and P_i inside the vesicles. For example, vesicle membranes are enriched in phosphatidylserine, a lipid having a high affinity for Ca^{2+}, as well as in the Ca^{2+}-binding proteins calbindin D9K and bone sialoprotein. Matrix vesicles are also rich in annexins, Ca^{2+} channels that export the ions from extracellular fluid into vesicles. P_i is provided by type-II Na/P_i cotransporters and by a specialized phosphatase (PHOSPHO1), which hydrolyzes phosphatidylcholine (PCho) and phosphatidylethanolamine (PEA) scavenged from the MV membrane (**Figure 4**).

The association between Ca^{2+} and P_i affords hydroxyapatite as $Ca_{10}(PO_4)_6(OH)_2$. When the concentration of this compound exceeds its solubility, the mineral precipitates inside vesicles as a fine deposit (**Figure 4**, step 1). This is followed by excretion of the nanocrystal seeds through the MV membrane into the extracellular fluid. There the crystals undergo further growth (step 2) supported by sufficient Ca^{2+} and P_i concentrations in the extracellular fluid (crystal growth requires lower concentrations of the ions than initiation of crystallization). The hydroxyapatite crystals then tightly associate with collagen fibrils (provided by osteoblasts) in the skeletal matrices, thus forming composite bone material (step 3). The P_i to inorganic pyrophosphate (PP_i) ratio controls the second step of mineralization since the latter inhibits crystal growth. PP_i is supplied by nucleotide pyrophosphatase, phosphodiesterase 1 (NPP1) from ATP hydrolysis [47]. In addition, PP_i is provided by ankylosis protein homolog (ANKH), which is an inorganic pyrophosphate transport regulator localized on the membranes of osteoblasts. Tissue-nonspecific alkaline phosphatase (TNAP) hydrolyzes PP_i to generate P_i, thus controlling the PP_i/P_i ratio (**Figure 4**).

Importantly, the HAP structure contains fully ionized $(PO_4)^{3-}$ species and an ionized hydroxyl group of water coordinated by one of the Ca^{2+} ions. High pKa values (12.37) for deprotonation of the $(HPO_4)^{2-}$ group, which would lead to $(PO_4)^{3-}$ and the pKa for ionization of Ca^{2+}-coordinated water (12.5) to generate OH^- ion, suggest that alkaline pH favors HAP formation. This is consistent with the observation that the optimal pH for osteoblast functioning is 8.4 [48]. Also, the alkaline pH is optimal for TNAP, which provides P_i for biomineralization in bone tissues. An alkaline environment is maintained by pumping H^+ ions from the medium into osteoblasts by chloride-proton antiporters.

3.3 Bone resorption

During an organism's life, bones undergo constant changes associated with initial bone growth and the need to rejuvenate bone through the removal of the aged-bone segments and replacement with new material (for review, see [46]). As alkaline conditions favor bone formation, bone destruction requires acidification, which converts HAP to water-soluble compounds. Osteoclasts control this process spatially and metabolically by recognizing aged-bone segments and creating an acidic environment between the cell and underlying bone (**Figure 5**). This triggers events causing the

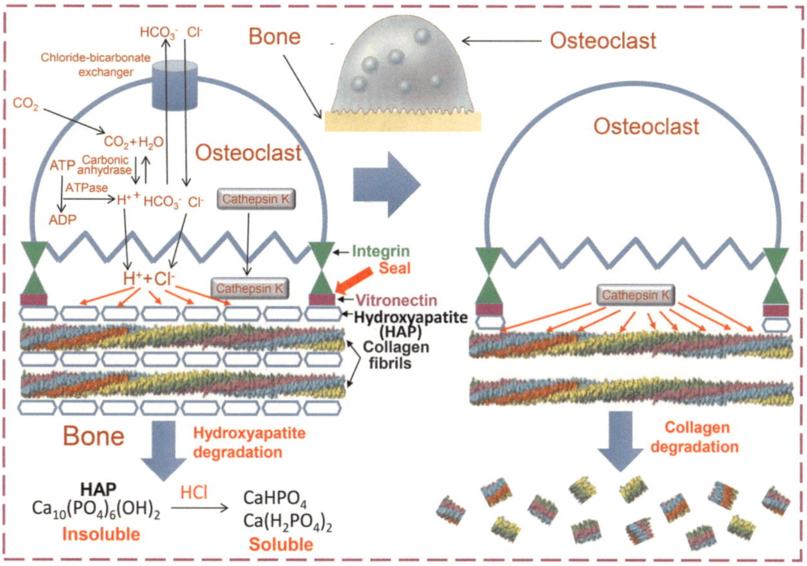

Figure 5.
The mechanism for osteoclast-mediated bone resorption. Osteoclast connects to the bone-forming isolated area between the cell and the bone followed by secretion of the bone-destructive agents (hydrochloric acid and cathepsin K) into the cavity. The secreted acid degrades HAP deposits, followed by proteolytic removal of the remaining collagen fibrils by cathepsin K. This is an extended and modified version of the original figure of [46].

formation of lysosomal vesicles that contain cathepsin K, a protease that finalizes bone destruction at later stages. Acidification is achieved through the action of carbonic anhydrase, which converts carbon dioxide to carbonic acid in a reversible reaction with water. Exchange of the resulting bicarbonate ion for the chloride anion of hydrochloric acid by a chloride-bicarbonate exchanger shifts the equilibrium of CO_2 hydration, thus enhancing acidification. Vacuolar H^+ ATPase generates additional acid through ATP hydrolysis. Diffusion of the resulting hydrochloric acid through the cellular membrane into isolated areas lowers the local pH to 4.5, resulting in protonation of the phosphate anions (pKa are 7.2 and 12.7) in the HAP structure, which produces more soluble Ca-phosphate forms, $CaHPO_4$ and $Ca(H_2PO_4)_2$. This causes the complete dissolving of mineralized HAP, thereby exposing the remaining organic collagen matrix to degradation by lysosomal-derived proteases, particularly cathepsin K. After degradation is complete, osteoclast moves to the adjacent bone segment to continue the degradation process. Notably, bone formation/resorption processes are finely balanced. Enhanced resorption rate causes bone loss disorders (e.g., osteoporosis), while extensive bone growth results in osteochondroma.

3.4 Pathological calcification in soft tissues

3.4.1 Arterial medial calcification (AMC)

Regulation of blood Ca^{2+} and P_i levels is a complex, precisely tuned process involving many tissues and hormones [49]. Dysregulation of this system can significantly affect bone and vasculature [50]. Chronic kidney disease (CKD), in particular, is associated with hyperphosphatemia and hypocalcemia promoting the development of arterial medial calcification (AMC) and impaired bone turnover [51]. Vascular calcification is the pathological deposition of calcium phosphate, often as hydroxyapatite, in the arteries and the heart valves. Originally, arterial medial calcification was considered a passive process caused solely by high serum levels of P_i and Ca^{2+}. However, it is currently accepted that the development of this disease is a complex cell-mediated process that is predominantly driven by pathological reprogramming of vascular smooth muscle cells, VSMCs [49].

Under normal conditions, the action of local and circulating inhibitors prevents pathological calcification [52, 53]. However, the imbalance between calcification inhibitors and calcification inducers can trigger vascular smooth muscle cell trans-differentiation, thus converting these cells into a species that resembles osteoblasts that execute physiological bone formation. The association of the same enzymes and ion transporters with both bone mineralization and arterial medial calcification supports this hypothesis. Thus, trans-differentiated vascular smooth muscle cells are able to generate extracellular vesicles that are functionally analogous to matrix vesicles produced by osteoblasts (compare **Figures 4** and **6**). Like matrix vesicles, extracellular vesicles concentrate Ca^{2+} and P_i and act as hatcheries for hydroxyapatite nanocrystals that escape into the arterial fluid where they grow into larger species. Grown crystals then deposit on various protein matrixes that include collagen and elastin, an abundant structural protein in the arterial wall. Crystals also deposit on cellular debris arising from extensive vascular smooth muscle cell apoptosis [46]. The latter is an important factor that contributes to arterial medial calcification. This phenomenon contrasts with bone production in which osteoblasts involved in hydroxyapatite deposition remain alive during the mineralization process.

Osteoblasts secrete large amounts of the collagenous matrix, which is then mineralized to form extended bone nodules; in contrast, calcifying vascular smooth muscle

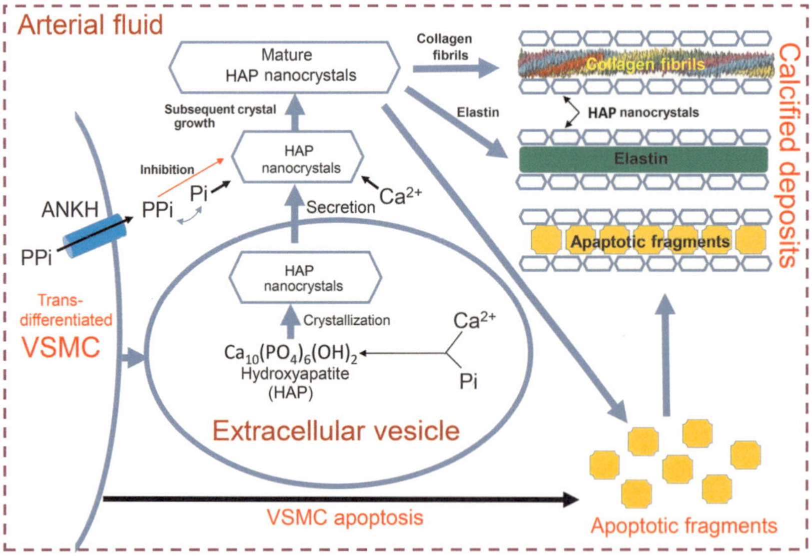

Figure 6.
*The mechanism for arterial medial calcification. Transdifferentiated vascular smooth muscle cells (VSMC) generate extracellular vesicles that produce HAP nanocrystals by mechanisms similar to that for matrix vesicles produced by osteoblasts (see **Figure 4**). These crystals are released into the arterial fluid, where they undergo further growth and then associate with various substrates to form calcified deposits. This is an extended and modified version of the original figure of [42].*

cells tend to form small, disordered but discrete regions of calcification, where hydroxyapatite deposits form on unstructured cellular debris [54].

As in the case of physiological hydroxyapatite deposition in bone tissues, Ca^{2+} and P_i are the major inducers of pathological calcification, while PP_i and ATP inhibit hydroxyapatite crystallization. However, there is a principal distinction between these two processes with regard to age-dependent dynamics. Thus, with age, the bone-producing capacity decreases, while pathological calcification intensifies [46].

Importantly, in both osteoblast and calcifying vascular smooth muscle cell cultures, the use of high phosphate concentrations ($\geq$ 5 mM) supersedes cell-mediated control of hydroxyapatite precipitation, resulting in widespread, nonspecific mineral deposition [43, 54]. This underscores a general paradigm of biological regulation in which the processes must proceed at moderate rates in order to be controlled.

3.4.2 Kidney stones

Kidney stones [55] are believed to have been following mankind since prehistoric times: The first established cases are dated to 4000 BC [56]. Kidney stones are the most common disease of the urinary tract. They have been associated with an increased risk of other serious diseases, including chronic kidney disease, end-stage renal failure, cardiovascular diseases, diabetes, and hypertension (for review, see [57].

Urinary stones consist of crystal matter impregnated with organic material (the matrix) that includes polymers, such as proteins, glycosaminoglycans, phospholipids, and carbohydrates. These molecules serve as a framework for kidney stone development. Albumin is the major component of the matrix of all stone types [58]. Along with calcium oxalate (CaOx), phosphate minerals (**Table 2**), brushite ($CaHPO_4{\cdot}2H_2O$), hydroxyapatite, $Ca_{10}(PO_4)_6(OH)_2$, and struvite ($Mg(NH_4)PO_4$) constitute a crystal matter in the majority of kidney stones cases [57].

Struvite represents a special case of kidney stones since its incidence is associated with a bacterial infection. It occurs among patients infected with urease-producing bacteria, such as *P. mirabilis, K. pneumonia*, *P. aeruginosa,* and *Enterobacter* [57, 59, 60]. Urease degrades urea to ammonia and CO_2, thus enabling struvite formation by providing NH_3 and alkalizing the medium, which facilitates the mineral deposition by deprotonating hydrophosphate ion [61].

Calcium deposits are the predominant renal stones, comprising about 80% of all urinary calculi [60]. The proportion of calcium stones may account for pure calcium oxalate (CaOx) (50%), hydroxyapatite (5%), and a mixture of both (45%). Urine acidification (pH of 5.0 to 6.5) favors CaOx stones, whereas calcium phosphate stones form when pH is greater than 7.5 [62]. This is attributed to deprotonation constants for oxalic (pKa_1 = 1.27 and pKa_2 = 4.28) and phosphoric acids (pKa_2 = 7.2 and pKa_3 = 12.7), as mineral formation requires Ox^{2-} and $(PO_4)^{3-}$ anions.

Kidney stone pathogenesis is a complex biochemical process that is not understood in detail. Renal stone formation involves physicochemical processes in supersaturated urine. Crystal deposition is affected by pH (see above), degree of supersaturation, and most importantly by the presence of numerous medium components that can either promote or impede precipitation. The maintenance of balance between crystallization promoters and inhibitors normally prevents precipitation. Therefore, kidney stone incidence is a combination of two major factors that contribute to supersaturation and dysregulation of inhibitors/promoters levels (**Figure 7**). The level of urinary saturation with respect to stone-forming constituents, such as calcium, phosphate, uric acid, oxalate, cystine, and low urine volume, is risk factor for crystallization [57].

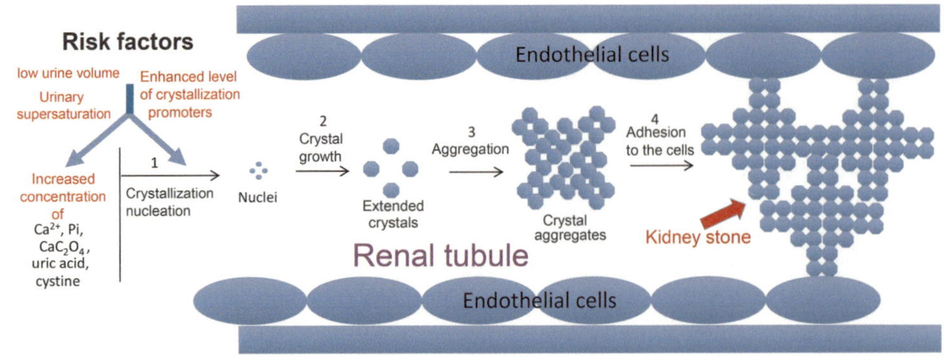

Figure 7.
The mechanism of kidney stone formation. Risk factors include: i) urea supersaturation with poorly soluble materials and/or ii) increased concentration of crystallization inducers, promote the formation of transient unstable clusters of nanocrystals (nuclei) (step 1). Nuclei can further grow to stable bigger crystals (steps 2 and 3), which then associate into large conglomerates (step 4) able to adhere to renal tube walls, and maturate into kidney stones. This figure is original.

All stones share similar events with respect to the mineral phase formation (**Figure** 7). They include nucleation, growth, and aggregation, followed by adhesion of the aggregates to renal tube walls that retain the crystals within the kidneys [63], thereby enabling their further growth therein.

At the nucleation step, stone-forming salts associate in solution into small, reversible, unstable clusters that may increase in size with the addition of new components [64]. In the kidney environment, nuclei usually form on existing surfaces in urine, following heterogeneous nucleation. Thus, epithelial cells, urinary casts, red blood cells, as well as crystals of various types can act as nucleating centers. The saturation extent necessary for heterogeneous nucleation is much lower than that for homogenous nucleation (see above), since when nuclei are created and anchored, they act as templates for the formation of the subsequent layers. That lowers the free energy of the process. Generally, the crystals grow slowly. Theoretical considerations suggest a low probability of a particle achieving a pathophysiologically relevant size by the sole process of crystal growth [60]. In this respect, the dramatic increase in the size of developing stones must be caused by the aggregation of existing crystals. This process involves the self-association of the original small crystals through surface-to-surface interaction. Loose aggregates are usually removed from the kidneys by urine flow. However, some of them can anchor to renal tubes, which retain the particle within the kidneys, thus enabling their further development into mature species.

3.5 Involvement of changes in inorganic phosphate physical state in regulation of transcription

Metabolic pathways controlled by enzymes are regulated by protein factors and small molecules (e.g., cellular metabolites) that interfere with or enhance enzyme activity. We found that transcription, the initial step of gene expression, is highly sensitive to changes in the physical state of inorganic phosphate in the surrounding medium [65]. Since this phenomenon represents a previously unknown regulatory factor and may play a new role in the cell, the author describe it in detail.

Transcription starts with recognition by RNA polymerase of the beginning of a gene, which is distinguished by a promoter sequence, followed by the melting of the

DNA double helix. This enables RNA synthesis, using one DNA strand as a template. Short RNA products synthesized in the course of transcription initiation are unstable and easily dissociate, forcing the enzyme to restart synthesis. The transition from initiation to the elongation step of RNA synthesis is accompanied by relinquishing the recognition of RNA polymerase contacts with the promoter and a dramatic stabilization of RNA in the transcription elongation complex (TEC). The TEC is highly processive, but it dissociates in response to a termination signal encoded in the DNA sequence at the end of the transcribed gene. Transcription regulation occurs mostly at the initiation step of RNA synthesis owing to the inherent instability of the RNA polymerase initiation complex.

We have shown that inorganic phosphate, in the cellular form of $MgHPO_4$, can exist *in vitro* at physiological concentrations in a supersaturated state in which the $MgHPO_4$ concentration in solution significantly exceeds (10- to 13-fold) its equilibrium solubility (**Figure 8A**), or as a precipitate. We discovered that *in vitro* $MgHPO_4$ in both of these distinct physical states strongly modulates RNAP activity. In the supersaturated metastable soluble form, $MgHPO_4$ stimulates the intrinsic RNA polymerase exoribonuclease activity (about 2000-fold) as described above (**Figure 2**), whereas as a precipitate, $MgHPO_4$ strongly and selectively inhibits initiation of RNA synthesis (**Figure 8B**).

At physiological pH, simple P_i salts are highly water-soluble. However, P_i in complex with magnesium ($MgHPO_4$) is poorly soluble (thermodynamic solubility ~1.5 mM). Considering the high intracellular concentrations of P_i and Mg^{2+} and their ~1 mM complex formation constant, the bulk of intracellular P_i must exist as a precipitated complex with Mg^{2+}. However, establishing the solubility equilibrium for $MgHPO_4$ requires prolonged incubation (**Figure 8A**); therefore, a metastable, supersaturated solution could persist in the cell. In addition, agents that inhibit $MgHPO_4$ crystallization could stabilize a supersaturated solution in the cell (analogous to hydroxyapatite deposition cases described above).

The effects of $MgHPO_4$ on transcription suggest that P_i could play a new major role in the regulation of gene expression. How it affects transcription *in vivo* would depend critically on its physical state in the cell, which has not yet been determined. Life is a nonequilibrium process with respect to cellular energetics and concentrations of

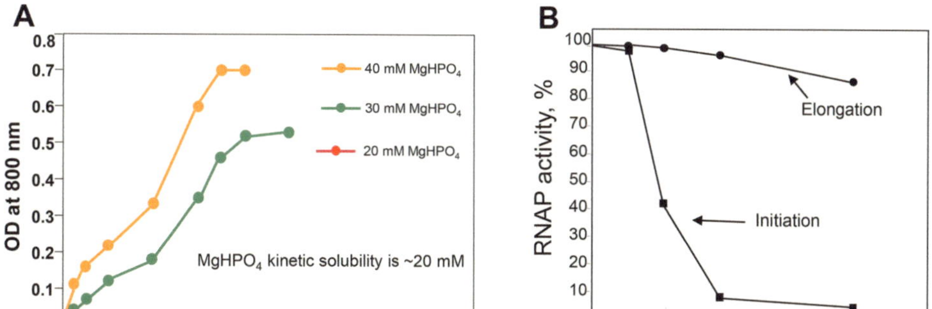

Figure 8.
Determination of the kinetic solubility for $MgHPO_4$ and effect of precipitated $MgHPO_4$ on transcription by RNA polymerase. A. Precipitation time course for $MgHPO_4$ generated by mixing the equivalent amounts of K_2HPO_4 and $MgCl_2$ at indicated concentrations. B. Effect of precipitated $MgHPO_4$ on RNAP polymerase activity in initiation and elongation steps of RNA synthesis. Charts are adopted from [65].

cellular metabolites [66]. Therefore, finding $MgHPO_4$ in a metastable, supersaturated state in the cell would contribute to understanding life processes by extending the concept of the nonequilibrium nature of life to the state of a particular cellular metabolite. Finding $MgHPO_4$ as a precipitate would strongly suggest that the compound participates in transcription regulation by affecting RNA synthesis through this novel mechanism.

3.6 Mineral deposition in nanobacteria

Nanobacteria gained broad attention in 1996, after a sensational announcement by President Bill Clinton that evidence for life on Mars, at least in the distant past, finally had been found. A meteorite originating from the surface of Mars about 15 million years ago and found in Antarctica appeared to contain the fossil remains of tiny bacteria-shaped life forms. Strikingly, similar formations have been found in Earth's minerals. Moreover, the ground-breaking discovery that these relict entities, named nanobacteria (NB) for their small size (40–500 nm), populate body fluids and organs of contemporary organisms and cause a number of diseases, added fuel to the excitement surrounding the phenomenon of nanobacteria. That prompted detailed investigations and the generation of far-reaching ideas that often seemed to outpace the facts. Due to unusual controversial properties, questions remained about what nanobacteria actually were and what they were not. After more than two decades, our understanding about the nanobacteria has advanced considerably. It appears that these "creatures" are neither exotic new pathogens nor new forms of life, even though they display astonishing life-like behavior. They turn out to be significant to human health and are believed to have played a formidable role in the origin and early evolution of life. Below is a brief description of the nanobacteria phenomenon and its impact on medical and nanomaterials research plus its implications for understanding the origin of life.

3.6.1 Discovery and properties of nanobacteria

In 1993, Robert L. Folk performed a systematic study of the rock collection from Italian hot springs [67], testing the idea that mineral deposition in the springs is mediated by bacteria. He described unusual mineral formations that he called "nannobacteria." The discovered entities, small spheres, morphologically resembled the fossilized remains of bacteria apparently possessing cell walls and filamentous structures (see **Figure 9**) [71, 72]. However, the observed particles were much smaller (from 10 to 200 nm) than common bacteria (1000–3000 nm). A few years later, similar structures were discovered by the McKay group in a Martian meteorite [71], whose age was estimated to be 4.5 billion years. Martian structures strongly resembled those found by Folk and were considered as evidence for the possibility of previous life on Mars and common principles guiding the creation of living entities in the solar system. These independent findings addressing the same paradigm were met with great enthusiasm. However, as attractive as they were, all the claims were based on morphological evidence alone. In addition, the observed particles seemed to be too small to support life as unicellular organisms. In the midst of the controversy, E. Olavi Kajander group found nanobacteria in biological materials, thus providing the first apparently direct evidence for nanobacteria as living organisms [73]. By examining commercial culture media, these researchers found small contaminating particles that

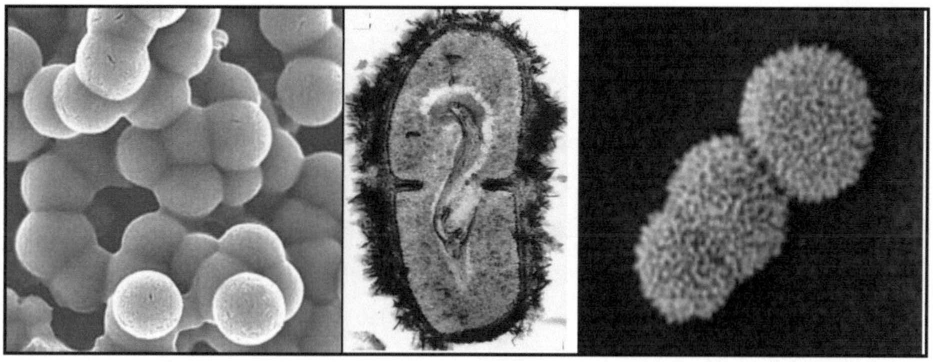

Figure 9.
Images of nanobacteria. Adopted from [68–70].

affected cellular growth and were resistant to all elimination techniques, including heat, detergent inactivation, irradiation, or antibiotic treatment. The observed particles ranged between 50 and 500 nm in size and possessed strong morphological similarity to those identified by Folk and McKay in Martian rock.

The major attributes of contemporary life forms are the presence of nucleic acids and proteins. These molecules were reported in nanobacteria [74]. DNA sequence analysis related these bacteria (which were named *Nanobacterium sanguineum)* to a subgroup of pathogenic bacteria that includes *Brucella* and *Bartonella*. The nanobacteria also displayed the unusual feature known as pleomorphism, the ability to change shape in response to growth conditions [74]. Thus, nanobacteria change from small spherical bodies to films and clumps of mineralized material. The mineral comprising these species turned out to be a common biological material—hydroxyapatite, HAP [68]. Surprisingly, subsequent studies suggested the ubiquitous nature of nanobacteria, finding them in most animal and human body fluids examined: blood, saliva, and urine, among others [68, 75]. Moreover, these entities were associated with abnormal mineral agglomerations (e.g., kidney stones [69]). Later, nanobacteria were linked to many other diseases, including various types of cancer, atherosclerosis, and degenerative diseases, such as arthritis, scleroderma, multiple sclerosis, peripheral neuropathy, Alzheimer's disease, and even viral infections, such as HIV (for review, see [75]). These reports promoted the idea that nanobacteria behave as transmissible pathogens and, as such, represent a serious health hazard [75].

Despite these findings, significant skepticism remained regarding the nature of nanobacteria. Thus, Jack Maniloff of the University of Rochester Medical Center labeled them "the cold fusion of microbiology." This is why many researchers called these entities "calcifying nanoparticles" (CNP) to emphasize their nonliving nature. Detailed structural and morphological studies established the ability of nanobacteria to self-propagate [74]. The features of self-propagation were highly sensitive to culture conditions, in particular to the presence of serum (**Figure 10**). Nanobacteria grow slowly (duplication time about 3 days compared to about 30 min for *Escherichia coli* and other commonly studied bacteria). In serum-containing media, nanobacteria grew in number and size by adding new HAP layers to their surface (**Figure 10**, stage 1). Growth continues through stage 2 in which the particles eventually acquire a dumbbell shape. These mature forms could be indefinitely (for years) passaged by dilution (stage 3) to produce the species of the same shape and size (50–400 nm).

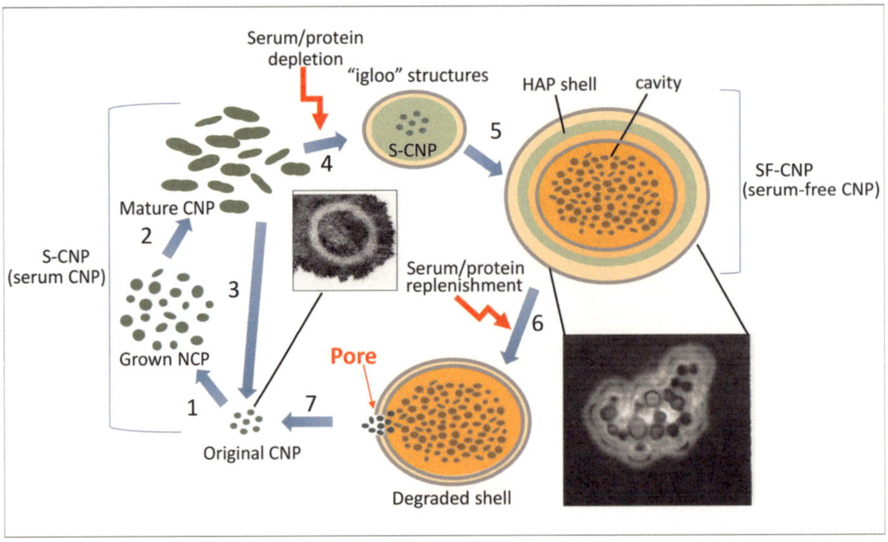

Figure 10.
"Life cycle" of nanobacteria under various conditions. In the presence of serum, nanobacteria grow and acquire dumbbell shape (stages 1 and 2). These species can be cultured indefinitely by serial dilutions (step 3). In the absence of serum, nanobacteria form multilayer extended structures with a cavity harboring small original species, which proliferate inside (stages 4 and 5). Subsequent addition of serum results in dissolution of the "igloo" outer shell (stage 6), releasing the small nanobacteria (stage 7), which can further proliferate in the presence of serum. The figure is a modified scheme of [74]. The photo images are adopted from [68].

These small coccobacillus-shaped calcifying nanoparticles (CNP) are referred to in **Figure 10** as serum calcifying nanoparticles (S-CNP). However, if these S-CNP are passaged in serum-/protein-free medium, the CNP produce biofilm-like structures in which they attach to the surface of the culturing vessel. They then increase significantly in size (1–10 μm) by acquiring several apatite mineral layers on their surface, forming "igloos" or "shells" (**Figure 8**, stages 4–6). Remarkably, the extended "igloo" forms harbor in their interior small S-CNP species (**Figure 10**), thus creating "dwelling places" and displaying a complex "life style" [74]. These igloo structures were named serum-free CNP (SF-CNP). Serum addition to SF-CNP cultures caused three major changes: i) detachment of SF-CNP from the culturing vessel surface, ii) degradation of the outer shell, followed by iii) release of the S-CNP from the SF-CNP interior through the developing pores. These changes are explained by action of serum components that promote hydroxyapatite shell solubilization. Normally, these components control hydroxyapatite crystallization upon CNP proliferation when culturing is performed in the presence of serum, thus maintaining the CNP small size.

3.6.2 Reconstruction of nanobacteria

Another milestone in the elucidation of the phenomenon of nanobacteria (NB) was work led by John O. Cisar [76], which provided the first reliable alternative view of NB as nonliving formations. Phospholipids, common components of cell membranes, bound to both calcium and phosphate in the medium, directed the formation of hydroxyapatite crystals. The small crystalline clumps seeded this way displayed a remarkable resemblance to nanobacteria in several respects. Like nanobacteria, these crystalline formations grew and replicated in culture as if they were alive. It was also pointed out that the presence of nucleic acid sequences that had been previously identified as hallmarks of nanobacteria could be accounted for by contamination commonly present in the laboratory environment, reagents, and glassware. This assumption seems reasonable, considering the high similarity of the presumed nanobacterial DNA to that of DNA from common bacteria, while significant differences in the DNA sequence are expected due to the exotic nature of nanobacteria. The proposal also explained the limited number of proteins found in nanobacteria, suggesting that these proteins are captured from the environment rather than being encoded in nanobacterial DNA. Collectively, the data strongly suggested that nanobacteria do not represent living entities, at least as they are commonly defined.

Shortly after Cisar's findings, the Kajander group described nanoparticles isolated from calcified blood vessels that not only retained DNA and proteins, but also were able to synthesize these molecules, as judged by the incorporation of radioactive precursors [77]. The measured rate of the DNA/protein synthesis was about 10,000-fold slower than in common bacteria. The exact mechanisms of these processes were not elucidated, leaving open the question about whether the observed activities are intrinsic to nanobacteria.

Following the work of John O. Cisar, several other groups undertook studies aimed at i) determining the fine structural organization of nanobacteria, and ii) dissection of the particles' chemical and biological nature to address the principal question of whether these entities are alive. Initial experiments set out to determine whether nanobacteria could be reconstituted from pure materials. The experiments used calcium phosphate (which is a mineral constituent of nanobacteria) and calcium carbonate, another compound with a natural tendency to aggregate in a precise molecular pattern to form crystals [78–80]. The deposition of these compounds was examined in the presence of various biological substances that inhibit crystal growth. It was expected that the addition of these substances would either delay or completely inhibit crystallization. Surprisingly, the mineral agglomerations continued to grow and propagate as nanoparticles, assuming shapes and geometries that made them look identical to nanobacteria. Moreover, nanoparticles grown under these conditions acquired cell-like walls and appeared to divide just like living bacteria. It was also found that the nanoparticles made of calcium carbonate-phosphate mixtures can bind to a variety of biological substances, including any charged molecule, small organic compounds (such as carbohydrates), lipids, or even DNA. Binding to charged groups stabilized the growing particles, promoting their further growth and modulating their shape.

Among the charged groups studied, the most interesting and complex effects were produced when crystallization was carried out in the presence of albumin or fetuin-A, proteins that are abundant in the blood and that possess a high affinity for Ca^{2+} ions [81]. Indeed, albumin alone accounts for half of the calcium-binding capacity of blood serum, while fetuin-A is even more unusual, capable of binding not only to calcium but also to calcium phosphate in the form of nascent apatite. Normally, these proteins inhibit calcium-phosphate deposition in soft tissues by scavenging Ca^{2+} and P_i ions from the environment. However, when the ion-binding capacity of these proteins is exceeded, they trigger hydroxyapatite deposition. As a result, these proteins become imbedded in the growing calcium nanoparticles. The studies concluded that the growing nanoparticles simply capture any readily available protein in their surrounding environment that is capable of binding Ca^{2+} and hydroxyapatite. In this view, the nanobacteria may be considered simple by-products of normal calcium metabolism in the absence of clearance mechanisms [70] that maintain optimal concentrations of the ions. This hypothesis explains previous data about the generation of antibodies against nanobacteria: The antibodies developed to proteins randomly captured on a nanoparticle. Indeed, it was shown that commercial antibodies sold as diagnostic tools for nanobacteria are in fact detecting fetuin-A and albumin. The discovery of DNA in nanobacteria reported in the early studies can be accounted for by the same effect.

Despite overwhelming evidence demonstrating that nanobacteria are nonliving particles that develop from biological minerals and organic materials, these structures seem to play an important role in biology and human health. Strikingly, when injected into healthy organisms, nanobacteria induce disorders causing abnormal calcification, which points to the "infectious" potential of these entities [75, 82], thus underscoring their astonishing resemblance to living cells. The significance of nanobacteria as a threat to public health was recognized by developing and commercializing a means to deal with the issue. Thus, established companies Nanobac OY (later Nanobac Pharmaceuticals) designed antibodies to detect nanobacteria in human tissues and provide medicines to treat NB "infections". Moreover, some common antibiotics were shown to inhibit nanobacteria proliferation. These drugs act by interfering with hydroxyapatite deposition upon nanobacteria growth rather than by blocking specific protein targets as they do in conventional microorganisms.

In spite of the extensive studies reviewed above, exactly how NB self-propagate remains unclear. Detailed investigation of this process is of paramount significance with respect to the possible involvement of the NB phenomenon in the origin of life.

3.6.3 Properties of nanobacteria implicated in the origin and early evolution of life

3.6.3.1 Morphology, physical, and physicochemical properties

The remarkable feature of nanobacteria particles is their ability to assemble from the surrounding mineral and organic components plus their broad promiscuity in choosing the building materials. Thus, the content of the nanoparticles can be easily modulated by changing the surrounding medium composition. Despite these factors that cause changes in shape and size, nanoparticles retain their striking resemblance to live biological forms, thereby reflecting morphological "stability" and the general nature of the nanobacterium phenomenon. This resilience and adaptability to changing growth conditions are remarkable properties that represent the hallmarks of life. In this regard, elucidation of the detailed mechanisms for nanoparticle formation from minerals and simple organic molecules under the conditions existing on Earth billions of years ago can aid in understanding how the first life forms evolved.

Biophysical studies suggest that nanobacteria-like particles represent a model for primitive biosystems [83] since they possess: i) small size, which allows for the refuge in the smallest crevices and pores in crustal rocks, thus elevating the chance of survival during geological events as extreme as major meteorite impacts; ii) a central cavity protected by an outer mineral shell against changing hostile environment and chemical factors; for example, treatment with 0.1 M HCl at 20°C for 1 day does not affect nanobacteria survival; iii) exceptional resistance to radioactivity, which could be a life-threatening factor on the early Earth; nanobacteria tolerate radiation doses as high as 15 kGy; iv) partial porosity and selective permeability of the mineral shell, which is a prerequisite for passage of liquids and cell signals; v) a pumping mechanism allowing transport of nutrients and organic compounds across the mineral shell into the central cavity allowing metabolic products to escape, as required for growth; vi) an ability to collect, organize, and store calcium, carbonate, and phosphate—essential inorganic components for primal biochemical processes; and vii) capability to use solar energy for activating, accelerating, and maintaining metabolic reactions and proliferation. The spherical shape of nanobacteria, combined with their physical properties, suggests that apatite nanoparticles could collect and concentrate light into their core, which affects NB growth. Indeed, cultured NB exposed to solar-intensity white light responded with an enhanced replication rate.

3.6.3.2 Catalytic properties of hydroxyapatite

Generally, catalysis includes the stimulation of a chemical process by lowering its activation barrier as the reactant molecules bind to a catalyst. The catalytic nature of all chemical reactions occurring in cells is a major feature of life. Therefore, it is thought that catalysis must have been involved in very early steps in the origin of life [84]. In current organisms, catalysis is mostly performed by protein, while during early life this function must have relied on minerals. Minerals possess remarkable catalytic potency. Below, the author review the catalytic properties of phosphate minerals with an emphasis on hydroxyapatite due to its likely role in life as a constituent of nanobacteria.

Catalysis occurs on a crystal surface that is exposed to reactant molecules in the surrounding medium and possesses functional groups that assist chemical conversions. In the inner layers of the crystals, atoms and ions are involved in interactions that maintain crystal structure. In the surface layer, only some of the functional groups of the crystal, those facing the interior of the crystal, are involved in interactions, while the groups exposed to the outer space retain "free" valences that can bind molecules from the surroundings and activate them for subsequent chemical reactions. **Figure 11**

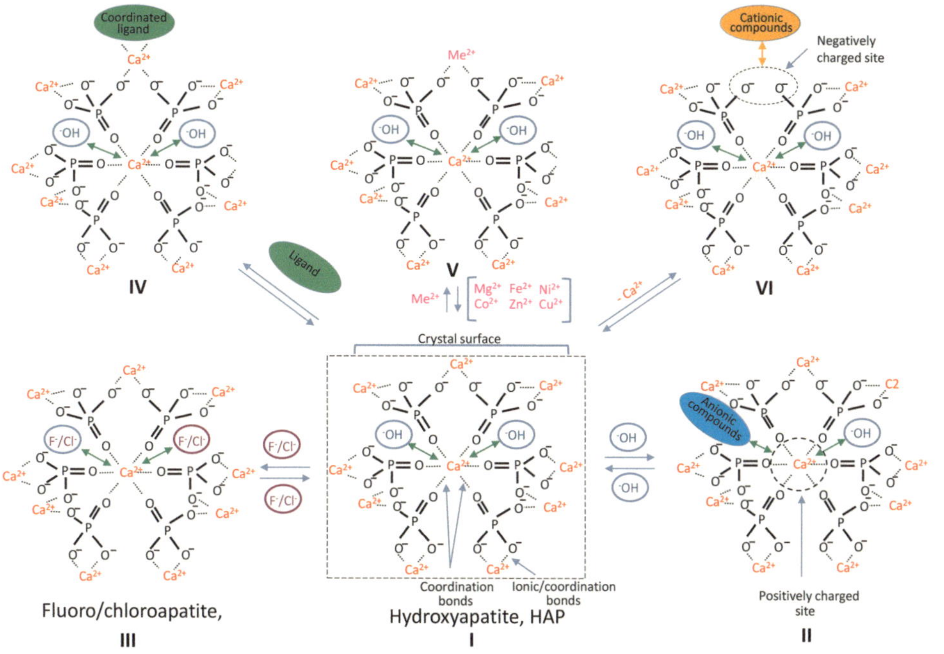

Figure 11.
Physicochemical processes that can occur on hydroxyapatite crystal surface. I. the structure of the HAP crystal unit. II. Coordinated hydroxyl groups can exchange for other anionic molecules (exemplified in III by fluoro/chloroapatite formation). Surface-exposed Ca^{++} ions can: i) bind ligands from solution through coordination (IV); ii) exchange for other di- and polyvalent metal ions (structure V), or iii) dissociate into surrounding medium creating a negatively charged center (VI) that can bind other cationic compounds. This figure is original.

depicts the structure for the crystal unit of hydroxyapatite (HAP), $Ca_{10}(PO_4)_6(OH)_2$ (structure I). This unit is assembled through ionic/coordination interactions of Ca^{2+}, OH^-, and P_i ions. Occasional dissociation of the hydroxyl groups or Ca^{2+} ions from the HAP crystal surface creates positively or negatively charged sites (structures II and VI, respectively), that are able to bind counter-ions from the solution. This is exemplified by fluoro- and chloroapatite, species that constitute a mineral component of teeth (structure III). Also, surface-exposed Ca^{2+} can coordinate ligands from the medium (structure IV). In addition, these Ca^{2+} ions can be easily exchanged for other metal ions (structure V), which modulates and diversifies the catalytic properties of the surface as described below. Remarkably, carbonate ions can incorporate into HAP structures (up to 6%), further contributing to the variety of possible interactions.

The ability of HAP to retain compounds through ionic and coordination interactions has been widely exploited by using HAP as a sorbent for chromatographic separation of numerous classes of biological compounds that include proteins, nucleic acids, polysaccharides, and small metabolites ([85] and refs. therein). In addition, coordinated Me^{2+} ions on the crystal surface can execute electrophilic catalysis (exemplified in **Figure 2**), which is performed by a wide variety of modern enzymes that have Me^{2+} ions at the active centers. In particular, this type of catalysis is displayed by enzymes involved in nucleic acid metabolism. Therefore, the HAP surface possesses enzyme-like properties, being able to bind reactant molecules and provide functional groups to assist chemical reactions as described below. **Table 3** summarizes the most important biologically relevant catalytic reactions that are stimulated by HAP.

Phosphorylation of organic molecules is the major reaction thought to be common in early life forms [1–4]: It would be the first process that combined organic and inorganic matter into the same entity. Anhydrides of phosphoric acid are considered to be the first phosphorylating agents. These energy-rich compounds are also plausible condensing agents for prebiotic biopolymer synthesis [94]. Therefore, various strategies have been tested in attempts to synthesize the simplest phosphoanhydride, pyrophosphate (PP_i), using the compounds and conditions that might have existed on early Earth. Strikingly, the HAP efficiently assists the synthesis of pyrophosphate (**Table 3**) from inorganic phosphate in the presence of cyanate [86], urea [87], or thioesters [88], simple compounds that were likely available in the primordial environment.

Phosphorylation generally results in the formation of phosphate esters with organic molecules. To understand how this process could be carried out in the primordial environment, investigations have focused on the chemistry and conditions affording reasonable yields of phosphorylation products. In the course of these studies, the HAP was found to efficiently catalyze nucleoside phosphorylation to yield monophosphate derivatives. At certain conditions, they underwent further phosphorylation, producing nucleoside di- and triphosphates [89–91], which are phosphoanhydride derivatives (**Table 3**). Importantly, this reaction provides substrates that can polymerize into polyribonucleotides that constituted the basis for the "RNA world," presumed to be very early life forms (before DNA).

Ribose is a building block of the nucleic acid backbone and a part of energy-rich compounds, NTPs that drive numerous life-supporting reactions in the cell. As such, this carbohydrate represents the most fundamental pentose. This monosaccharide might have been created from C1 or C2 carbon sources in a prebiotic environment. For example, formaldehyde (C1) was found in the dust within the coma of a comet, and glycolaldehyde (C2, a dimer of formaldehyde) was found in the gas around a

Reaction	Catalyst	References
Pyrophosphate synthesis		
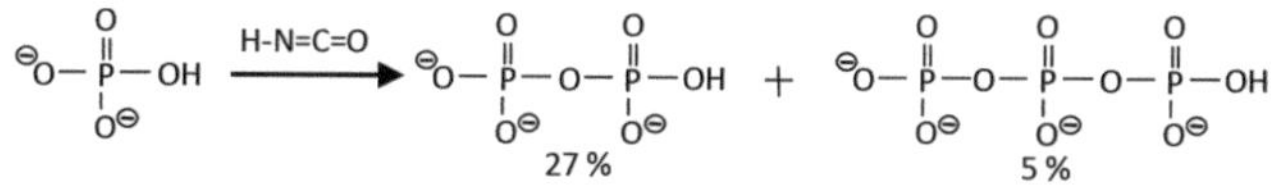	HAP	[86]
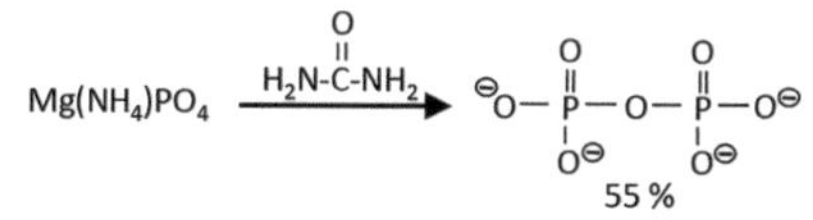	HAP, struvite	[87]
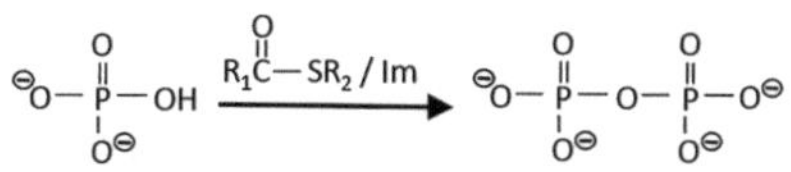	HAP	[88]
Nucleoside and nucleotide phosphorylation		
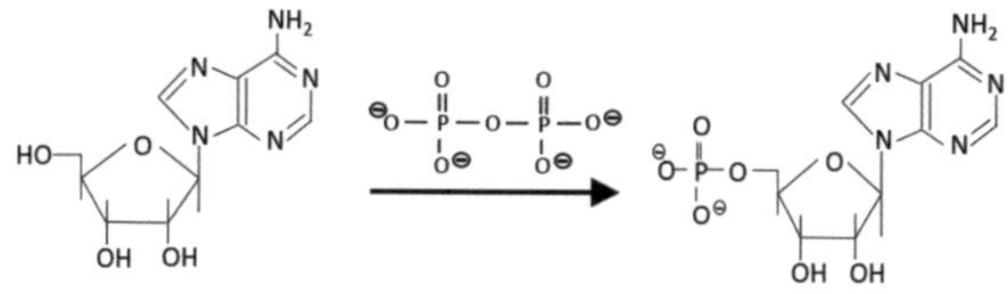	HAP	[89–91]
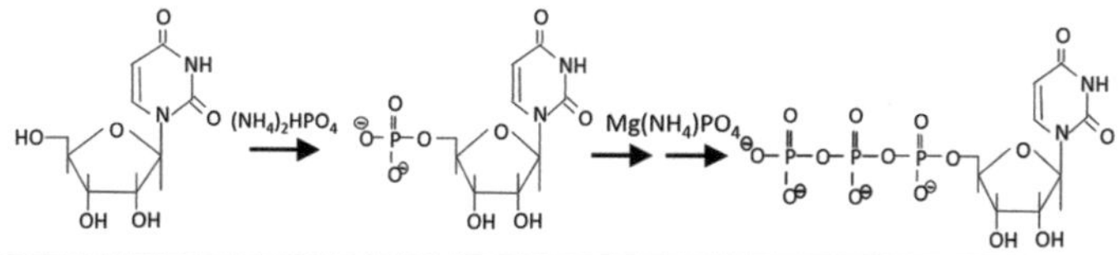	HAP, struvite	[87]
Ribose synthesis		
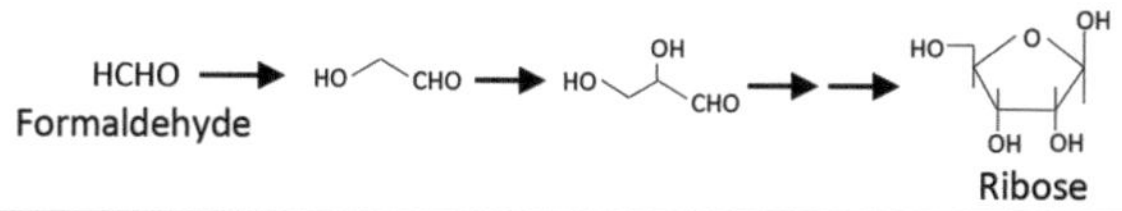	HAP	[92]
Template-directed RNA synthesis		
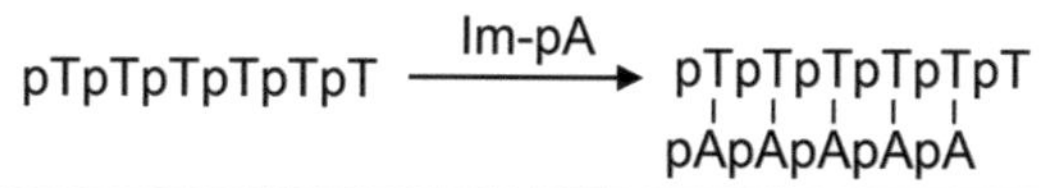	HAP	[93]

Table 3.
Biologically significant reactions catalyzed by phosphate minerals.

forming star. One of the possible synthetic routes to ribose is the "formose reaction" pathway (**Table 3**) discovered by Aleksandr Butlerow in 1861 [95]. In several attempts at reactions from formaldehyde and glycolaldehyde, strongly alkaline aqueous solutions, containing calcium ions, were required and complex mixtures of reaction products, including tetroses, pentoses, and hexoses formed in this process. Strikingly, using hydroxyapatite as a catalyst allowed the production of ribose from C1 and C2 carbon sources in one pot of hot water [92]. This catalyst worked continuously in multistep reactions to produce ribose preferentially rather than the other monosaccharides.

Finally, the ability of HAP to assist RNA synthesis from activated monomeric building blocks (**Table 3**) has been explored [93]. Polyuridylic acid, poly(U), is adsorbed completely from aqueous solution by hydroxyapatite under conditions that permit template-directed synthesis of RNA oligoadenylates in free solution. Even though the yield of oligoadenylates was enhanced to almost the same extent by poly (U) in the presence or the absence of hydroxyapatite, hydroxyapatite adsorbed higher molecular-weight oligoadenylates selectively from a mixture of oligomers. Based on these results a mechanism for prebiotic oligonucleotide formation has been proposed in which selective adsorption of oligomeric RNA products on HAP provides a protective environment, while monomers are released from the surface for reactivation/recycling.

It should be noted that HAP modification achieved by changing the surface-exposed Ca^{2+} ions to other polyvalent metal ions (as shown in **Figure 11**, structure V) greatly extends the repertoire of chemical reactions that can be assisted by this mineral [96]. This finding underscores a remarkable analogy with contemporary protein-based metalloenzymes, whose catalytic properties can be modulated by changing the active center-bound metal ions ([97] and references therein).

3.6.3.3 Possible involvement of nanobacteria in the origin of life and early evolution

Life-like behavior of apatite-based nanobacteria particles, coupled with remarkable catalytic properties displayed by hydroxyapatite in biologically significant reactions, invites an attractive hypothesis about how these nanoparticles evolved into life entities.

As pointed above, the formation of nanobacteria requires an organic substrate on which hydroxyapatite deposits. Therefore, nanoparticle formation requires the initial production of organic material. This material could be generated in reactions occurring spontaneously in the environment and accumulate in the "primordial soup" [98, 99]. As suggested, these reactions were likely promoted by light, electric discharge, high temperature, and/or the catalytic action of various minerals. Interaction of hydroxyapatite with the accumulated organic material could produce nanobacteria-like structures. Based on the properties and the complex nanobacteria behavior described above, these structures could be considered as models for, if not the first, mostly inorganic primitive life forms in which a hydroxyapatite shell served as both a catalytic entity assisting chemical reactions and a membrane secluding the pool of synthesized molecules. A subsequent step in evolution could be accompanied by the increased involvement of the nanobacteria-produced organic material in the nanoparticles' composition and functioning. This organic material could be generated through catalytic processing of the simple organic molecules absorbed on the mineral surface of nanobacteria from the environment followed by transport of the synthesized material into the nanoparticle's interior using a pumping mechanism [83]. This would i) provide protection of the produced organic material from environmental perturbations and ii) retain and concentrate the material in the interior. The accumulated matter could undergo further transformation into more complex and diverse "libraries" of compounds assisted by the interior hydroxyapatite surface of the shell acting as a general "enzyme" possessing multiple catalytic activities and broad substrate specificity. This transformation would be facilitated by the elevated concentration of the molecules in the cavity relative to that in the surrounding medium. Some of the synthesized molecules could be used as cofactors in HAP-catalyzed processes, thus increasing the repertoire and specificity of catalyzed chemical reactions.

At this stage, primitive cells could start undergoing a natural selection process with respect to their fitness. Thus, small differences in the structure and chemical composition of the accumulated organic material could affect the "growth," so that faster proliferating nanobacteria would dominate the environment. The preservation of the features of the selected cells would be ensured by cell division creating identical species.

Subsequent steps of nanobacteria evolution could include the production of RNA, which in its self-duplicating form would constitute the basis of the proposed "RNA world" [29–32]. This step could be followed by creation of DNA and proteins through establishing a genetic code, by relating nucleic acid sequence to that of an encoded protein, thereby setting metabolic pathways resembling those in contemporary organisms. This process would produce catalytically active protein structures—primordial enzymes that would offer more efficient catalysis and greater substrate specificity than hydroxyapatite. However, the majority of the enzymes would still depend highly on metals—invariant catalytic entities and constituents of minerals. This is supported by the fact that about a third of contemporary enzymes require metal for their function [100, 101]. Remarkably, in ribozymes (RNA-based enzymes) catalytic centers include metal ions coordinated by phosphate residues—structures reminiscent of hydroxyapatite. In addition, some contemporary enzymes possess mineral nanoclusters (e.g., iron sulfide, pyrite [102], or iron sulfide-molybdenum nanoclusters [103]), suggesting that even in modern life forms minerals cannot be completely substituted by protein structures.

Later evolution could include the production of lipid-like compounds, which due to their high affinity to HAP (see above) could pave the outer surface of HAP shells, thereby forming membranes and creating a species of greater resemblance to modern cells. The final step could involve a fundamental change in cell lifestyle by eliminating the need for hydroxyapatite structures as catalysts, cell membranes, and cell division-driving entities.

3.7 Why nature has chosen magnesium as a major metal ion to support life processes

As discussed above, phosphate minerals are believed to play an exceptional role in the origin and evolution of early life by executing catalytic reactions and by providing inorganic phosphate, a crucial material for emerging life forms. Another major player in this event is magnesium, since Mg^{2+} ion is a ubiquitous cofactor assisting virtually all cellular reactions that include the transfer of phosphoryl or nucleotidyl groups [104]. The high prevalence of Mg^{2+}-mediated catalysis in the biochemistry of phosphometabolites strongly suggests that this metal, along with P_i, played an exclusive role in prebiotic chemistry. Thus, magnesium-assisted reactions involving phosphate compounds must have strongly contributed to a pool of primordial "metabolites"—building blocks for first life forms [104]. In this regard, the high content of the principal inorganic ions (Mg^{2+} and P_i) in the environment must have been crucial factors defining the rates of chemical reactions involved in life. Most of the polyvalent metal cations in the Earth's interior are balanced by polyvalent inorganic anions to form poorly soluble minerals, thus excluding them from life processes. However, this balance is not absolute. Thus, in nature, Mg^{2+} and Ca^{2+} ions seem to be in excess of precipitating polyanions. This explains the relatively high concentration (50 mM and 10 mM, respectively) of the ions in seawater, whose composition reflects the dissolution of the Earth material over the span of billions of years. Whereas Ca^{2+} and Mg^{2+} ions are abundant, the concentration of P_i ions in seawater is very low (about 1–3 μM). This is the result of a general imbalance of P_i anion versus polyvalent metal cations, which scavenge the inorganic phosphate into insoluble minerals (e.g., HAP, whose solubility is about 1 μM).

An intriguing question is about concentrations of the major inorganic ions (Mg^{2+} and P_i) that were required to maintain a reasonable rate of chemical reactions involved in the origin of life. Reconstitution of the environments where life emerged is obviated by the paucity of geological records. However, some insight can be gained from theoretical considerations. According to a well-accepted paradigm, "chemistry conservation principle" [105, 106], the chemical composition of major inorganic ions of organisms is more conservative than the changing environment and, as such, retains information about the ancient environment in which the first cells emerged. In other words, the intracellular concentration of major inorganic ions in current life forms is close to that in the environment in which the protocells came into existence. In this view, the P_i and Mg^{2+} ion concentration in the habitats of the first cells is expected to be 1–10 mM. While Mg^{2+} concentration in the seawater is consistent with this paradigm, the P_i concentration therein is about 1000-fold lower, strongly suggesting that life originated in localized "puddles" (where P_i content could be higher due to proximity to a soluble P_i source), rather than in extended water reservoirs (e.g., oceans). The same conclusion was reached in the studies [107] considering the other life-supporting inorganic ions. Importantly, as suggested by the "chemistry conservation principle" the local environment where life emerged should be distinguished by low Ca^{2+} content (about 0.1–1 μM), which is a concentration of Ca^{2+} ions in contemporary cells. This could be rationalized in terms of a scavenging effect of P_i and bicarbonate that precipitate Ca^{2+} ions as HAP and calcite, respectively. However, the presence of moderate Ca^{2+} concentrations in the primordial media would still allow the formation and functioning of nanobacterial structures, which are presumed to have an important role in the origin of life.

So, what minerals could supply the high concentrations of the major ions needed for prebiotic chemistry? The most soluble P_i form (**Table 2**) is represented by the natural mineral newberyite ($MgHPO_4 \cdot 3H_2O$), whose solubility (about 1.5 mM) is in the range of that of Mg^{2+} and P_i ion concentrations in the cell. Therefore, this mineral could be a source for both ions in concentrations necessary to originate life. The higher temperature of the ancient environment (55–85°C) could further increase the mineral solubility and therefore the concentration of the ions in water reservoirs. It is thought that newberyite could have been a source of condensed phosphates that executed prebiotic phosphorylation [108]. The significantly greater solubility of $MgHPO_4$ relative to phosphates of other divalent metal ions explains why nature has chosen Mg^{2+} as the major ion for electrophilic biocatalysis, as deduced in our previous study [97]. The high solubility of newberyite can be explained by the generally moderate coordination strength of Mg^{2+} ions as they establish bonds with oxy-ions of P_i ($K_d \sim 1.5$ mM), as compared to other Me^{2+} ions (e.g., transition metals). This property facilitates the magnesium mineral dissolution that is accompanied by destroying the coordination bonds, and it impedes precipitation, which requires coordination bond formation. In other words, Mg^{2+} and P_i ions are biocompatible, as they maintain high solubility when present in the same solution. Moreover, Mg^{2+} is compatible with another biologically significant inorganic ion, bicarbonate, as evidenced by the high solubility of $MgCO_3$ (about 5 mM) compared to that for bicarbonates of other divalent metals.

Notably, an additional mechanism could provide a reasonable rate of prebiotic phosphorylation, even at low P_i concentration in the medium, and it has been shown that HAP crystals could act not only as catalysts in phosphorylation of organic molecules, but also as a P_i source [86]. Therefore, a reasonable phosphorylation rate, in this case, is achieved by proximation catalysis as an organic molecule binds nearby the HAP-associated P_i at the crystal surface.

The above considerations define conditions that allowed life to originate and evolve, at least in the form existing on Earth, which is a fine balance of chemical elements and their forms that provide the soluble source of the most significant biological Mg^{2+} and P_i (and possibly $(HCO_3)^-$) ions. Thus, an excessive content of elements forming anions that precipitate Mg^{2+} ions would eliminate this key catalytic metal ion from the aqueous media. Likewise, the prevalence of polyvalent metal ions forming insoluble salts with inorganic phosphate and bicarbonate would exclude these vital anions from the media. Both processes would deplete from media key inorganic ions that are crucial for life.

3.8 Why nature has chosen phosphate as a part of life material

The phosphate choice for the construction of biological molecules has been widely discussed [1–3]. This question can be viewed from two perspectives. The choice could be determined by the versatile beneficial properties of P_i that are addressed above. On the other hand, it could be asked whether nature had a choice. In other words, can other elements be utilized in major biological structures as a phosphorus substitute? This idea inspired numerous studies in which arsenate $(AsO_4)^{3-}$ and vanadate $(VO_4)^{3-}$ anions act like phosphate in some biological processes (for review, see Refs. [109–111] and our results on RNA polymerase, presented in **Figure 2**). Arsenate affects about 200 enzymes in the cell by acting as a P_i analog [112]. This behavior can be explained by the similarity in physicochemical properties of these ions and P_i. Thus, geometry and dissociation constants for arsenate ($pKa_1 = 2.22$, $pKa_2 = 6.98$, and $pKa_3 = 11.53$) and vanadate ($pKa_1 = 3.5$, $pKa_2 = 7.8$, and $pKa_3 = 12.5$) ions are close to those for phosphate ($pKa_1 = 2.1$, $pKa_2 = 7.2$, and $pKa_3 = 12.7$). These studies culminated in sensational reports [113, 114] stating that DNA isolated from bacteria populating arsenate-rich waters of Mono Lake in California USA contained this structure, suggesting in particular that phosphate linkages in nucleic acids can be substituted by arsenate counterparts. However, subsequent studies failed to confirm this finding. Indeed, arsenate and vanadate esters and diesters are not stable and, as such, could not maintain nucleic acid structure or substitute for phosphate in other ester-based biological molecules. Thus, the half-life for arsenate esters in water solution is a fraction of a second [115]. Anhydrides of arsenate and vanadate are unstable as well, which would preclude the utilization of these compounds in cell energetics. Unlike phosphorus (which is a nonmetal) As is a metalloid, while V is a metal. This factor fundamentally affects the bond between the element and oxy- or alkoxy-groups. While the P-O bond is covalent and very stable under physiological conditions, As-O and V-O bonds are mostly coordination bonds, which are unstable. That allows a quick exchange between the bound groups and surrounding water. Therefore, it seems that nature did not have an alternative to phosphate. Although nonphosphate life forms cannot be excluded, they would be fundamentally different from those existing on Earth. However, with respect to forming hard tissues (e.g., bones), phosphate is dispensable, since carbonate and silicate minerals can be deposited to produce robust biomaterials, such as shells in mollusks [116] and membrane structures in diatoms [117].

Finally, there is a remarkable similarity in the chemistry of carboxylic and phosphoric acids, which is evidenced by using the same agents for the synthesis of major biologically relevant derivatives of these acids including anhydrides, esters, amides, and thioesters. This similarity could facilitate the inclusion of phosphate in the biosynthesis of metabolites in primitive life forms.

Acknowledgements

Dedicated to my father Abdullah Mustaev, in recognition of his achievements in the chemistry of rare-earth metal phosphates. The author is grateful to Karl Drlica, Max Gottesman, and Maxim Kozlov for kind support and help in manuscript preparation.

Author details

Arkady Mustaev
Department of Microbiology and Molecular Genetics, New Jersey Medical School, Public Health Research Institute, Rutgers Biomedical and Health Sciences, Newark, New Jersey, USA

*Address all correspondence to: arkantrien@yahoo.com

References

[1] Westtheimer FH. Why nature choose phosphate. Science. 1977;**235**:1173-1178

[2] Kamerlin S, Sharma P, Prasad R, Warshel A. Why nature really chose phosphate. Quarterly Reviews of Biophysics. 2013;**46**:1-132

[3] Hunter T. Why nature chose phosphate to modify proteins. Philosophical Transactions of Royal Society B. 2012;**367**:2513-2516

[4] Pasek MA, Kee TP. On the origin of phosphorylated biomolecules. In: Egel R, Lankenau D-D, Mulkidjanian AY, editors. Origins of Life: The Primal Self-Organization. Berlin/Heidelberg, Germany: Springer-Verlag; 2011. pp. 57-84

[5] Gull M. Prebiotic phosphorylation reactions on early earth. Challenges. 2014;**5**:193-212

[6] Ruiz-Merazo K, Briones., de la Escosura A. Prebiotic systems chemistry: New perspectives for the origin of life. Chemical Reviews. 2014;**114**:285-366

[7] Wolfenden R, Ridgeway C, Young G. Spontaneous hydrolysis of ionized phosphate monoesters and diesters and the proficiencies of the phosphatases and phosphodiesterases as catalysts. Journal of the American Chemical Society. 1998; **120**:833-834

[8] Lipkin D, Preston T, Talbert PT, Cohn M. The mechanism of the alkaline hydrolysis of ribonucleic acids. Journal of the American Chemical Society. 1954; **76**:2871-2872

[9] Kawamura K. Measurement of the rate of RNA hydrolysis in aqueous solution at elevated temperatures using a new monitoring method for hydrothermal reactions. Nucleic Acids Symposium Series. 1999;**42**:289-290

[10] Garcia AK, Schopf AK, Yokobori JW, Akanuma S, Yamagishi SA. Reconstructed ancestral enzymes suggest long-term cooling of Earth's photic zone since the Archean. Proceedings of the National Acadamey of Sciences USA. 2017;**114**: 4619-4624

[11] Ray A, Norden B. Peptide nucleic acid (PNA): Its medical and biochemical applications and promise for the future. The FASEB Journal. 2000;**14**: 1041-1060

[12] Dunn J, Grider MH. Physiology, Adenosine Triphosphate. Treasure Island (FL): StatPearls Publishing; 2022

[13] Rao NN, Gómez-García MR, Kornberg A. Inorganic Polyphosphate: Essential for Growth and Survival. Annual Review of Biochemistry. 2009; **78**:605-647

[14] Kusano S, Ishihama A. Functional interaction of Escherichia coli RNA polymerase with inorganic polyphosphate. Genes to Cells. 1997;**2**: 433-441

[15] Agledal L, Niere M, Ziegler M. The phosphate makes a difference: Cellular functions of NADP. Redox Report. 2010; **15**:2-10

[16] Su-Ju Lin S-J, Guarente L. Nicotinamide adenine dinucleotide, a metabolic regulator of transcription, longevity and disease. Current Opinion in Cell Biology. 2003;**15**:241-246

[17] Zimmerman SB, Little JW, Oshinsky CK, Gellert M. Enzymatic joining of DNA strands: a novel reaction of diphosphopyridine nucleotide. Proceedings of the National Acadamey of Sciences USA. 1967;**57**:1841-1848

[18] Clark JF. Creatine and phosphocreatine: A review of their use in exercise and sport. Journal of Athletic Training. 1997;**32**:45-50

[19] Ibba M, Dieter Soll D. Aminoacyl-tRNA synthesis. Annual Review of Biochemistry. 2000;**69**:617-650

[20] Kotra LP, Golemi D, Amro NA, Liu G-Y, Mobashery S. (1999) Dynamics of the lipopolysaccharide assembly on the surface of Escherichia coli. Journal of the American Chemical Society. 1999;**121**: 8707-8711

[21] Mandell DJ, Chorny I, Groban ES, Wong SE, Levine E, Rapp CS, et al. Strengths of hydrogen bonds involving phosphorylated amino acid side chains. Journal of the American Chemical Society. 2007;**129**:820-827

[22] Holt LJ. Regulatory modules: Coupling protein stability to phopshoregulation during cell division. FEBS Letters. 2012;**586**:2773-2777

[23] Knouse KW, Flood DT, Vantourout JC, Schmidt MA, Mcdonald IM, Eastgate MD, et al. Nature chose phosphates and chemists should too: How emerging P (V) methods can augment existing strategies. ACS Central Science. 2021;7: 1473-1485

[24] Gottesman ME, Mustaev A. Inorganic phosphate, arsenate, and vanadate enhance exonuclease transcript cleavage by RNA polymerase by 2000-fold. Proceedings of the National Academy of Sciences of the United States of America. 2018;**115**:2746-2751

[25] Sosunov V, Sosunova E, Mustaev A, Bass I, Nikiforov V, Goldfarb A. Unified two-metal mechanism of RNA synthesis and degradation by RNA polymerase. The EMBO Journal. 2003; **22**:2234-2244

[26] Sosunov V, Zorov S, Sosunova E, Nikolaev A, Zakeyeva I, Bass I, et al. The involvement of the aspartate triad of the active center in all catalytic activities of multisubunit RNA polymerase. Nucleic Acids Research. 2005;**33**:4202-4211

[27] Kettenberger H, Armache KJ, Cramer P. Architecture of the RNA polymerase II-TFIIS complex and implications for mRNA cleavage. Cell. 2003;**114**:347-357

[28] Steitz T. A mechanism for all polymerases. Nature. 1998;**391**:231-232

[29] Ekland EH, Bartel DP. RNA-catalysed RNA polymerization using nucleoside triphosphates. Nature. 1996; **382**:373-376

[30] Johnston WK, Unrau PJ, Lawrence MS, Glasner ME, Bartel DP. RNA-catalyzed RNA polymerization: Accurate and general RNA templated primer extension. Science. 2001;**292**:1319-1325

[31] Zaher HS, Unrau PJ. Selection of an improved RNA polymerase ribozyme with superior extension and fidelity. RNA. 2007;**13**:1017-1026

[32] Wochner A, Attwater J, Coulson A, Holliger P. Ribozyme catalyzed transcription of an active ribozyme. Science. 2011;**332**:209-212

[33] Dorozhkin SV, Epple M. (2002) biological and medical significance of calcium phosphates. Angewandte Chemie, International Edition. 2002;**41**: 3130-3146

[34] Watt JC. The behavior of calcium phosphate and calcium carbonate (bone salts) precipitated in various media, with applications to bone formation. Biology Bulletin. 1923;**44**:280-317

[35] https://en.wikipedia.org/wiki/Solubility_table

[36] https://www.chm.uri.edu/weuler/chm112/refmater/KspTable.html

[37] Verbeeck CRMH, De Bruyne PAM, Driessens FCM, Verbeek F. Solubility of magnesium hydrogen phosphate trihydrate and ion-pair formation in the system magnesium hydroxide-phosphoric acid-water at 25 degree. Inorganic Chemistry. 1984;**23**(13):1922-1926

[38] https://www.fishersci.com/shop/products/magnesium-hydrogen-phosphate-trihydrate-98-extra-pure-thermo-scientific/AC264780010

[39] Ofoegbu SU. Technological challenges of phosphorus removal in high-phosphorus ores: Sustainability and possibilities for greener ore processing. Sustainability. 2019;**11**:6787. DOI: 10.3390/su11236787

[40] Prakash KH, Kumar R, Ooi CP, Cheang P, Khor KA. Apparent solubility of hydroxyapatite in aqueous medium and its influence on the morphology of Nanocrystallites with precipitation temperature. Langmuir. 2006;**22**:11002-11008

[41] Aage HK, Andersen BL, Blom A, et al. The solubility of struvite. Journal of Radioanalytical and Nuclear Chemistry. 1997;**223**:213-215

[42] Bourne LE, Wheeler-Jones C, PD., Orriss IR. Regulation of mineralization in bone and vascular tissue: A comparative review. The Journal of Endocrinology. 2021;**248**:R51-R65

[43] Hortells L, Sosa C, Millan Á, Sorribas V. Critical parameters of the in vitro method of vascular smooth muscle cell calcification. PLoS One. 2015;**10**:e0141751

[44] Orimo H. The mechanism of mineralization and the role of alkaline phosphatase in health and disease. Journal of Nippon Medical School. 2010;**77**:4-12

[45] Bone strength. (https://phys.libretexts.org/Bookshelves/Conceptual_Physics/Book%3A_Body_Physics_-_Motion_to_Metabolism_(Davis)/07%3A_Strength_and_Elasticity_of_the_Body/7.01%3A_Strength_of_Human_Bones)

[46] Ross FP, Christiano AM. Nothing but skin and bone. The Journal of Clinical Investigation. 2006;**116**:1140-1149

[47] Vaingankar SV, Fitzpatrick TA, Johnson K, Goding JW, Maurice M, Terkeltaub R. Subcellular targeting and function of osteoblast nucleotide pyrophosphatase phosphodiesterase. American Journal of Physiology. Cell Physiology. 2004;**286**:C1177-C1187

[48] Galow A-M, Rebl A, Koczan D, Gimsa J. MC3T3 osteoblast-like cells cultured at alkaline pH: Microarray data (Affymetrix gene Chip mouse 2.0 ST). Data in Brief. 2017;**13**:108-114

[49] Shanahan CM, Crouthamel MH, Kapustin A, Giachelli CM. Arterial calcification in chronic kidney disease: Key roles for calcium and phosphate. Circulation Research. 2011;**109**:697-711

[50] Giachelli CM. Vascular calcification mechanisms. Journal of American Society of Nephrology. 2004;**15**:2959-2964

[51] Yamada S, Giachelli CM. Vascular calcification in CKD-MBD: Roles for phosphate, FGF23, and klotho. Bone. 2017;**100**:87-93

[52] Zhu D, Mackenzie NC, Farquharson C, Macrae VE. Mechanisms and clinical consequences of vascular calcification. Frontiers in Endocrinology. 2012;**3**:95

[53] Orriss IR. Extracellular pyrophosphate: The body's 'water softener'. Bone. 2020;**134**:115243

[54] Patel JJ, Bourne LE, Davies BK, Arnett TR, Macrae VE, Wheeler-Jones CP, et al. Differing calcification processes in cultured vascular smooth muscle cells and osteoblasts. Experimental Cell Research. 2019;**380**: 100-113

[55] Giannossi L, Summa V. A review of pathological biomineral analysis techniques and classilcation schemes. In: Aydinalp C, editor. An Introduction to the Study of Mineralogy. London, UK, InTech, IMAA-CNR, Italy: InTechOpen; 2012

[56] Lopez M, Hoppe B. History, epidemiology and regional diversities of urolithiasis. Pediatric Nephrology. 2008; **25**:49-59

[57] Alelign T, Petros B. Kidney stone disease: An update on current concepts. Advances in Urology. 2018;**2018**: 3068365. DOI: 10.1155/2018/3068365

[58] Khan SR, Kok DJ. (2004) modulators of urinary stone formation. Frontiers in Bioscience. 2004;**9**:1450-1482

[59] Barbasa C, Garciaa A, L., Saavedraa, L., Muros, M. Urinary analysis of nephrolithiasis markers. Journal of Chromatography B. 2002;**781**:433-455

[60] Coe FL, Evan A, Worcester E. Kidney stone disease. The Journal of Clinical Investigation. 2005;**115**: 2598-2608

[61] Grffith DP. Struvite stones. Kidney International. 1978;**13**:372-382

[62] Kumar SBN, Kumar KG, Srinivasa V, Bilal S. A review on urolithiasis. Internatioanl Journal of University of Pharmacy and Life Sciences. 2012;**2**: 69-280

[63] Chhiber N, Sharma M, Kaur T, Singla S. Mineralization in health and mechanism of kidney stone formation. International Journal of Pharmaceutical Science and Invention. 2014;**3**:5-31

[64] Chaudhary A, Singla SK, Tandon C, C. In vitro evaluation of Terminalia arjuna on calcium phosphate and calcium oxalate crystallization. Indian Journal of Pharmaceutical Sciences. 2010;**72**:340-345

[65] Gottesman ME, Mustaev A. Change in inorganic phosphate physical state can regulate transcription. Transcription. 2019;**10**:187-194

[66] Ornes S. Core concept: How nonequilibrium thermodynamics speaks to the mystery of life. Proceedings of the National Academy of Sciences of the United States of America. 2017;**114**: 423-424

[67] Folk RL. SEM imaging of bacteria and nannobacteria in carbonate sediments and rocks. Journal of Sedimentary Research. 1993;**63**:990-999

[68] Kajander EO, Ciftcioglu N. Nanobacteria: An alternative mechanism for pathogenic intra- and extracellular calcification and stone formation. Proceedings of the National Acadamey of Sciences USA. 1998;**95**:8274-8279

[69] Ciftcioglu N, Bjorklund M, Kuorikoski K, Bergstrom K, Kajander EO. Nanobacteria: An infectious cause for kidney stone formation. Kidney International. 1999;**56**:1893-1898

[70] Young JD, Martel J. The rise and fall of nanobacteria. Sci Am. 2010;**302**:52-59. DOI: 10.1038/scientificamerican0110-52

[71] McKay DS, Gibson EK, Thomas-Keprta KL, Vali H, Romanek CS, Clemett SI, et al. Search for past life on Mars: Possible relic biogenic activity in martian meteorite ALH 84001. Science. 1996;**273**: 924-926

[72] Folk RL, Taylor LA. Nannobacterial alteration of pyroxenes in martian meteorite Allan Hills 84001. Meteoritics & PlanetaryScience. 2002;**37**:1057-1069

[73] Ciftcioglu N, Kuronen I, Åkerman K, Hiltunen E, Laukkanen J, Kajander EO. A new potential threat in antigen and antibody products: Nanobacteria. In: Brown F, Burton D, Doherty P, Mekalanos J, Norrby E, editors. Vaccines 97. Cold Spring Harbor, NY: Cold Spring Harbor Lab. Press; 1997. pp. 99-103

[74] Mathew G, McKay DS, Ciftcioglu N. Do blood-borne calcifying nanoparticles self-propagate? International Journal of Nanomedicine. 2008;**3**:265-275

[75] Kajander EO. Nanobacteria – Propagating calcifying nanoparticles. Letters in Applied Microbiology. 2006; **42**:549-552

[76] Cisar JO, Xu D-Q, Thompson J, Swaim W, Hu L, Kopecko DJ. An alternative interpretation of nanobacteria-induced biomineralization. Proceedings of the National Acadamey of Sciences USA. 2000;**97**:11511-11515

[77] Ciftcioglu N, Pelttari A, Kajander EO. Extraordinary growth phases of nanobacteria isolated from mammalian blood. Instruments, Methods, and Missions for the Investigation of Extraterrestrial Microorganisms. 1997; **3111**:429-435. SPIE

[78] Jan Martel J, Young JD-E. Purported nanobacteria in human blood as calcium carbonate nanoparticles. Proceedings of the National Acadamey of Sciences USA. 2008;**105**:5549-5554

[79] Wu CY, Young L, Young D, Martel J, Young JD. Bions: A family of biomimetic Mineralo-organic complexes derived from biological fluids. PLoS One. 2013;**8**: e75501

[80] García-Ruiz JM, Emilio Melero-García E, Stephen T, Hyde ST. Morphogenesis of self-assembled Nanocrystalline materials of barium carbonate and silica. Science. 2009;**323**: 362-365

[81] Raoult D, Drancourt M, Azza S, Nappez C, Guieu R, Rolain J-M, et al. Nanobacteria Are Mineralo Fetuin Complexes. PLoS Pathogens. 2008;**4**:e4

[82] Aoki H, Aoki H. Acute toxicity of hydroxyapatite microcrystal suspension by intravenous injection in rats. In: Transactions of the Annual Meeting of the Society for Biomaterials in Conjugation with the International Biomaterials. St Louis Park: Society for Biomaterials; 1996. p. 357

[83] Sommer AP, McKay DS, Ciftcioglu N, Oron U, Mester AR, Kajander EO. Living Nanovesicles - chemical and physical survival strategies of primordial biosystems. Journal of Proteome Research. 2003;**2**:441-443

[84] Muchowska KB, Varma SJ, Moran J. Nonenzymatic metabolic reactions and Life's origins. Chemical Reviews. 2020; **12**:7708-7744

[85] Itoh D, Yoshimoto N, Yamamoto S. Retention mechanism of proteins in hydroxyapatite chromatography – Multimodal interaction based protein separations: A model study. Current Protein & Peptide Science. 2019;**20**:75-81

[86] Miller SL, Parris. Synthesis of pyrophosphate under primitive earth conditions. Nature. 1964;**204**:1248-1250

[87] Handschuh GJ, Lohrmann R, Orgel LE. The effect of Mg 2+ and Ca 2+ on urea- catalyzed phosphorylation reactions. Journal of Molecular Evolution. 1973;**2**:251-262

[88] Weber RL. Formation of pyrophosphate on hydroxyapatite with thioesters as condensing agents. Biosystems. 1982;**15**:183-189

[89] Lohrmann R, Orgel LE. Urea-inorganic phosphate mixtures as prebiotic phosphorylating agents. Science. 1971;**171**:490-494

[90] Neuman MW, Neuman WF, Lane K. On the possible role of crystals in the origins of life. III. The phosphorylation of adenosine to AMP by apatite. Currents in Modern Biology. 1970;**3**: 253-259

[91] Neuman MW, Neuman WF, Lane K. On the possible role of crystals in the origins of life. IV. The phosphorylation of nucleotides. Currents in Modern Biology. 1970;**3**:277-283

[92] Usamia K, Okamoto A. Hydroxyapatite: Catalyst for a one-pot pentose formation org. Biomolecular Chemistry. 2017;**15**:8888-8893

[93] Gibbs D, Lohrman R, Orgel LE. Template-directed synthesis and selective adsorption of oligoadenylates on hydroxylapatite. Journal of Molecular Evolution. 1980;**15**:347-354

[94] Hulshof J, Ponnamperuma C. Prebiotic condensation reactions in an aqueous medium: A review of condensing agents. Origins of Life and Evolution of the Biosphere. 1976;**7**: 197-224

[95] Butlerow A. Bildung einer zuckerartigen Substanz durch Synthese. Justus Liebigs Annalen der Chemie. 1861;**120**:295-298

[96] Fihri A, Len C, Varma RS, Solhy A. Hydroxyapatite: A review of syntheses, structure and applications in heterogeneous catalysis. Coordination Chemistry Reviews. 2017;**347**:48-76

[97] Gottesman ME, Chudaev M, Mustaev A. Key features of magnesium that underpin its role as the major ion for electrophilic biocatalysis. The FEBS J. 2020;**287**:5439-5463

[98] Oparin AI, A.I. Proiskhozhdenie zhizny. Moscow. Izd: Moskovski Rabochi; 1924

[99] Haldane JBS. The Origin of Life. Rationalist Annual. 1929;**148**:3-10

[100] Silva J, Williams R. The Biological Chemistry of the Elements. Oxford: Clarendon Press; 1991

[101] Glusker J, Katz E, Bock C. Metal ions in biological systems. Rigaku Journal. 1999;**16**:8-16

[102] Beinert H. Iron-sulfur proteins: Ancient structures, still full of surprises. Journal of Biological Inorganic Chemistry. 2000;**5**(2–15):103

[103] Čorić I, Holland PL. Insight into the iron–molybdenum cofactor of Nitrogenase from synthetic iron complexes with sulfur, carbon, and hydride ligands. Journal of the American Chemical Society. 2016;**138**:7200-7211

[104] Holm NG. The significance of Mg in prebiotic geochemistry. Geobiology. 2012;**10**:269-279

[105] Macallum AB. The paleochemistry of the body fluids and tissues. Physiological Reviews. 1926;**6**:316-357

[106] Mulkidjanian AY, Galperin MY. Physico-chemical and evolutionary constraints for the formation and selection of first biopolymers: Towards the consensus paradigm of the abiogenic origin of life. Chemistry & Biodiversity. 2007;**4**:2003-2015

[107] Mulkidjanian AY, Bychkov AY, Dibrova DV, Galperin MV, Koonin EV. Origin of first cells at terrestrial, anoxic geothermal fields. Proceedings of the National Acadamey of Sciences USA. 2012;**109**:E821-E830

[108] Sales BC, Chakoumakos BC, Boatner LA, Ramey JO. Structural properties of the amorphous phases produced by heating crystalline $MgHPO_4{\cdot}3H_20$. Journal of Non-Crystalline Solids. 1993;**159**:121-139

[109] Crans DC, Smee JJ, Gaidamauskas E, Yang L. The chemistry and biochemistry of vanadium and the biological activities exerted by vanadium compounds. Chemical Reviews. 2004; **104**:849-902

[110] Németi B, Gregus Z. Mechanism of thiol-supported arsenate reduction mediated by phosphorolytic-arsenolytic enzymes: I. The role of arsenolysis. Toxicology Sciences. 2009;**110**:270-281

[111] Krebs HA, Eggleston LV, Knivett VA. Arsenolysis and phosphorolysis of citrulline in mammalian liver. The Biochemical Journal. 1955;**59**:185-193

[112] Ratnaike RN. Acute and chronic arsenic toxicity. Postgraduate Medical Journal. 2003;**79**:391-396

[113] Wolfe-Simon F, Blum JS, Kulp TR, Gordon GW, Hoeft SE, Pett-Ridge J, et al. A bacterium that can grow by using arsenic instead of phosphorus. Science. 2011;**332**:1163-1166

[114] Elias M, Wellner A, Goldin-Azulay K, Chabriere E, Vorholt JA, Erb TJ, et al. The molecular basis of phosphate discrimination in arsenate-rich environments. Nature. 2012;**491**:134-137

[115] Fekry MI, Tipton PA, Gates KS. Kinetic consequences of replacing the internucleotide phosphorus atoms in DNA with arsenic. ACS Chemical Biology. 2011;**6**:127-130

[116] Addadi L, Joester D, Nudelman F, Weiner S. Mollusk Shell formation: A source of new concepts for understanding biomineralization processes. Chemistry - A European Journal. 2006;**12**:980-987

[117] Hildebrand M, Lerch SJL, Shrestha RP. Understanding diatom Cell Wall Silicification—Moving forward. Frontiers in Marine Science. 2018;**5**:125. DOI: 10.3389/fmars.2018.00125

Chapter 3

Phosphorous Paradox and the Unsuspected Intrinsic Property of Human Beings to Dissociate the Water Molecule

Arturo Solís Herrera, María del Carmen Arias Esparza and Martha Patricia Solís Arias

Abstract

Phosphorous paradox means that this element is abundant on Earth, it is present inside of every cell of living things. However, is so scarce in the Universe. Phosphate, the most exploited form of phosphorous, is a vital constituent of fertilizer. Phosphate rock has emerged as a globally traded commodity linked to a diverse set of politically charged debates, ranging from environmental degradation and threats to human health to food security and agricultural sovereignty. Supposedly, life can multiply until all the phosphorus is gone, and then there is an inexorable halt, which nothing can prevent (Asimov, Isaac). Phosphorus seems like a Life's Bottleneck. It is so believed that Phosphorous (P) has been placed as a critical resource for the bio-economy and for food security at the global scale. The biogeochemical P flow has been described as a "planetary boundary," which, in parts of the world, has already been exceeded. However, our discovery about the unexpected intrinsic capacity of living beings to dissociate the water molecule breaks the ground. Thereby, the formation of Phosphorous requires the presence of Life.

Keywords: eutrophication, fertilizer, plant nutrition, hydrogen, nitrogen, water

1. Introduction

The purpose of this work is to concatenate the biochemical logic of the relatively recently discovered property of living entities to dissociate the molecule from water with the phosphorus paradox, as well as to present a novel method to efficiently manage the earth and water problems secondary to excess of phosphates in different bodies of water, based on the human eye's biology.

Theoretically, for phosphorus, there is no substitute, there is no element that can replace it [1]. Alfalfa can germinate and grow in agricultural soil containing 0.1% phosphorus, while the plant only contains 0.7% phosphorus in its structure. The structure/activity ratio of phosphorus makes it an important and irreplaceable element for plant growth. To date, there is no known way—natural or synthetic—that can carry out the functions that phosphorus performs. Curiously, in breast-fed infants, the phosphorus such as iron intake is very low [2].

Few centuries ago, phosphorus was chemically identified; however, throughout history, phosphorus has been used in the form of crop residues and manure that were dispersed in agricultural fields. This ancient practice continues so far, but an increase in phosphorus mining throughout the twentieth century contributed, at least initially, to steadily rising agricultural yields, but in the long term, the fertility of agricultural soil is adversely affected. Fertilizers manufactured with high proportions of phosphorus, nitrogen, and potassium boost the plant growth to unprecedented levels, especially in tropical soils that are poor in these constituents [3] although for some reason, nature so provides, and the proof is that these fertilizers, in the long run, contribute to impoverish yields.

In the 1960s, manufactured fertilizer was gearing up farmers to feed more people than the world had ever known; thus, harvests were ahead of a growing population. and although the number of people with malnutrition has decreased, the current figure of 925 million remains worrying [4].

Global production of phosphate rock is now nearly 13 times what it was in 1930s [5]. It has virtues as a key elemental the biochemical of life, but also phosphorus has also earned a well-deserved reputation as a persistent pollutant. In rural areas, unfortunately, phosphates regularly flow into receiving water as runoff from "fertilized" agricultural fields, [6] and in urban areas from sewage sources as a major constituent of human excreta flushed down toilets, as a result of indiscriminate use of phosphates as additives in industrialized food and drinks. Phosphorus can excessively boost local nutrient levels, promoting abnormal algal blooms in the lakes and rivers where it concentrates—a process called eutrophication [7].

Supposedly, this excessive algal growth can eventually lower oxygen levels in the water to the point where some fish species can no longer survive. But the reality is quite the opposite, as algal blooms are triggered precisely by low levels of dissolved oxygen caused in turn by high levels of phosphates. On the other hand, low levels of dissolved oxygen tend to affect marine species until eventually they disappear, regardless of whether there is an overpopulation of algae.

2. After all, for the prokaryotic and eukaryotic cell, water is not indivisible

The human eye has 3–4 ml of water that is not rechanged during all the life span of the individual. To practical aims, this is stagnant water; however, this water has adequate dissolved oxygen levels and rarely goes on acidity. We found the biochemical mechanism that Mother Nature uses to maintain the physicochemical characteristics of this eye's stagnant water in good shape for decades [8].

Melanin splits something previously thought to be unsplittable, and we'll never look at light, water, Universe, human being, and living things in the same way. The dissociation of the water molecule has transcendent industrial applications, some of them are exemplified in **Figures 1–12**.

Our finding that glucose—and thereafter meals in general—is just the building block of human being but not source of energy because light can be absorbed directly by living things, which suppose their capacity to transform light power into chemical energy in a previously unimaginable split form—like plants; means substantive advances in the fundamental understanding of light and how it behaves inside living things. Thereby, human body is not exception [9].

Figure 1.
QBLOCK™, a novel material developed based on human eye's Biology, which also dissociates the water molecules. At left, the container with the presence of QBLOCKS™ explains the abundance of bubbles. The container at right has no QBLOCK™, thereby it has no "bubbles" of oxygen. To date, 5 months later, the bubbles remain in the container with QBLOCK™.

Figure 2.
In experiments where QBLOCK™ is applied to offshore sand, after some months, this soil can support plants to grow up. Photograph shows sprouts of a mango seed after 8 days.

Figure 3.
After 2 weeks, the sprouts of mango seed. The QBLOCK™ was placed earlier, 14 months ago, and is deep in the offshore sand. Mango seeds were placed 2 weeks ago.

Figure 4.
Sprouts of mango seed after 16 days. Notice the QBLOCK™ on the surface.

Figure 5.
Mango and avocado sprouts.

Figure 6.
Avocado sprout in a soil treated con QBLOCK™.

Figure 7.
Mango sprouts, after 21 days. Notice the QBLOCK™ on the soil surface (and deep too).

Figure 8.
The presence of QBLOCK™ even in offshore sand, allowed mango seeds germination (right). At center avocado, right: mango sprouts. QBLOCK™ was used in the three specimens.

This is a major paradigm change of how we understand the interaction between light and living things in a way that was not believed to be possible. Not only did we find a new biochemical reaction entity, but it was one that nobody believed could exist [10].

Figure 9.
Left: mango. Right: tamarindo.

Figure 10.
The presence of larvae (yellow arrow) in water contaminated with carbon is unusual, due to formaldehyde formation (CH_2O), which is toxic. But in the presence of our QBLOCK™, the history is different, beginning with the rise of dissolved oxygen levels, which support life.

Phosphorus, a 5A element with atomic weight of 31, comprises just over 0.6% of the composition by weight of plants and animals [11]. A ubiquitous mineral on Earth, but not in the universe, and the second most abundant mineral in the human body, phosphorus represents ~1% of total body weight [12]. Common chemical linkage is

Figure 11.
In the flask with residues of grains (peanut), but with no other carbon source, only chlorophyll developed on the upper surface of QBLOCK™.

Figure 12.
In a closed bottle, simulating a closed system, the presence of QBLOCK™ immersed in the soil sample allows the development of organic carbon. The bottle has been closed since 2016, and a moderate amount of water was added one time only: at the very beginning of the experiment. Photograph taken in January 2022.

in the form of phosphate ester and phosphoanhydride. The element phosphorus is a key element in organic molecules overall in those involved in a wide variety of critical cellular functions. These include the biochemical temperature regulation through the hydrolysis of adenosine triphosphate (ATP), maintenance of genetic information with nucleotides DNA and RNA, intracellular signaling via cyclic adenosine monophosphate (cAMP), and membrane structural integrity via glycerophospholipids. But we keep in mind that energy, defined as everything that produces change, is not only needed to function, to move, reproduce, think, etc., but it is also required even to preserve the shape, the form, and in the case of the molecules and membranes, to continue as such.

The metabolism of inorganic phosphorus (Pi) is acting as a weak acid. At physiological pH of 7.4, Pi exists as both $H_2PO_4^{(-)}$ and $HPO_4^{(2-)}$ and acts as an extracellular fluid (ECF) buffer. Pi is the form transported across tissue compartments and cells. Eighty percent of the body phosphorus is present in the form of calcium phosphate crystals (apatite) that confer hardness to the bone and teeth and function as the major phosphorus reservoir. The remainder is present in soft tissues and ECF. The phosphorus coming from meals and liquids, comprising both inorganic and organic forms, is digested in the upper gastrointestinal tract.

During growth, there is net accretion of phosphorus, and with aging, net loss of phosphorus occurs, in similar way and at the same rate that the capacity of our body to dissociate the water molecule declines. Kidney is the main regulator of ECF Pi concentration by virtue of having a tubular maximum resorptive capacity for Pi (TmPi) that is under close endocrine control that means energy expenditure. It is also the main excretory pathway for Pi surplus, which is passed in urine, thereby requiring energy coming from water dissociation. At a dietary phosphorus of 1400 mg, 1120 mg is absorbed (energy required) in the upper intestine to the ECF, 210 mg returned to the intestine by endogenous secretion, processes all that need energy, resulting in 910 mg net Pi absorption and 490 mg fecal excretion. In the bone, 180 mg is deposited by bone formation and 180 mg return to the ECF by bone resorption, and all involved processes need power. In the kidney, 5040 mg is filtered at the glomerulus and 4130 mg return to the ECF by tubular reabsorption with 910 mg excreted in the urine. In soft tissue, Pi is exchanged between ECF and cells. Let us remember that any chemical reaction, any process, no matter how small, requires energy.

Bioavailability of phosphorus also varies depending on the source. Plant protein sources generally have the lowest bioavailability, followed by animal protein sources, then inorganic phosphate additives with the highest bioavailability. Inorganic phosphates have been nearly 100% bioavailable [13]. Phosphorus from plant-based sources remains less bioavailable than animal sources and animal sources less bioavailable compared with inorganic phosphate additives.

The growing Phosphorous "paradox" (the simultaneous overabundance of P impairing water quality and the prospect of global scarcity of P for future agricultural production) has stimulated new convergence between P-security and water-quality research agenda [14]. In the Universe, phosphorous is notably scarce.

Both in agricultural and urban systems, the fragmentation of the P cycle has implications even for water-quality impairment [15]. A sufficient (adequate for the purpose) and efficient (performing with the least waste of effort) utilization of P may offer a great reduction potential in animal husbandry and crop production [16].

The management of animals plays a key role in reducing P inputs to soils and, consequently, P losses from arable lands and grasslands. Because of the regional concentration of animal husbandry, improved diets with less P content may be most urgently required and effective in regions with high stocking density.

Genes involved in pathways relevant for P utilization were differentially expressed due to variable P supply. Phosphorous fluxes through various process and ecosystems along which originally mined and processed P is diluted and distributed over increasingly large parts of the terrestrial and aquatic environments.

Animals fed with low-P diets showed attempts to maintain mineral homoeostasis via intrinsic mechanisms [17]. Pyrophosphatase, an enzyme, could completely exchange the oxygen atoms within the phosphate ion with oxygen atoms originally within water molecules [18]; however, water does not release oxygen for free, so it is necessary to dissociate the molecule (of water) first. The isotopic composition of oxygen (proportion of Oxygen 16 to Oxygen 18) within phosphate ions in plant leaves was different from that observed in the solution delivering P to the plant [19], which suggests that inside living things, atomic nucleus tends to grow, but it does not happen in a flask solution. The difference is the available energy inside the living thing can impel protons (H^+) with enough and adequate force, so the atomic nucleus of oxygen and other elements increase their atomic number.

With the adequate surroundings, and a precise and enough energy, it is possible to create the main elements of periodic table contained in living things. Supposedly, only a synchrotron can impel subatomic particles with enough speed to be included in atomic nuclei. However, Nature can do it inside cells of living things.

For instance, Hydrogen has an atomic number of 1 and atomic mass of 1. Thereby Hydrogen has no neutrons. The difference between Carbon (6), Nitrogen (7), and Oxygen (8) is one and two protons, respectively.

The internal cellular environment has a relative abundance of Hydrogen and Oxygen (and energy) coming from water dissociation. Therefore, we can enlist Hydrogen, Carbon, Nitrogen, Oxygen, Sodium, Magnesium, Phosphorous, Sulfur, Chlorine, and Calcium as primary elements normally present in living things.

Now, we'll enlist the same elements with the number of protons in their atomic nuclei. H (1), C (6), Nitrogen (7), O (8), Na (11), Mg (12), P (15), Sulfur (16), Cl (17), K (19), and Ca (20).

Notice that the difference between them is just one proton, in general terms. Thereby, the formation inside living things of Carbon con atomic mass of 14, and Oxygen con atomic mass of 18, means that atomic nuclei can grow inside living cells. After death, Carbon 14, and Oxygen 18, tends to fade gradually along thousands of years.

This is, the growth of atomic nuclei of Carbon and Oxygen can occur inside the cell, because water dissociation happens there, so protons and energy, the two key elements, are available. Furthermore, these transformations—in guarded proportions—can also occur between Na (11) and Mg (12; between P (15) and S (16); also, between Cl (17), K (19), and Ca (20).

It is possible that a certain degree of transmutation between these elements happens to adjust the requirements of the living beings, allowing them a better adaptation to their surroundings. Thereby, Life is not totally dependent on determined diet, it can hatch under diverse diet composition, because the human body can synthesize chemical elements—guarded proportions—and compensate for nutritional deficiencies in the environment at any given time.

In relation to the essential trace elements, we have Manganese (^{25}Mn), Iron (^{25}Fe), Cobalt (^{26}Cu), Nickel (^{27}Ni), Copper (^{29}Cu), and Zinc (^{30}Zn) as examples of chemical elements whose difference between them is a proton, that is: a hydrogen without electron, and the dissociation of water produces them—electrons—abundantly [20].

Supplements are not the answer, i.e.: the intake of calcium supplements in patients with osteoporosis makes their bones brittle.

2.1 Phosphorous

Remarkable abundance of phosphorous on Earth and its scarcity in the Universe suggest strongly that the P could be formed by living beings overall those in soil. Phosphorus is widely distributed in the global food supply, with milk and dairy being the greatest contributors followed by meat and poultry. Notice a strong relationship with living beings. Circadian fluctuations in some bioactive components and trace elements are suggested to transfer chronobiological information from mother to child to assist the development of the biological clock [21].

For dairy cows, mineral P supplementation of the feed is generally not necessary and might be needed only when fed with high amounts of corn products. This includes more precise prediction of the dietary P requirement and a better characterization of the availability of different P sources used in animal feed [22].

About nonruminants, much attention has recently been given to the variation in plant P sources, in particular phytate-P [23]. It has been known for about two decades that the use of the enzyme phytase as a feed additive can effectively increase phytate-P availability in pigs and poultry. However, enzymes do not make possible an impossible reaction.

Sophisticated analytical techniques such as stable isotope techniques (33-P, 18-O), NMR- and synchrotron-based spectroscopies are required for quantifying P cycles, fluxes, and dynamics in the soil and other environmental systems.

Measurement of the isotopic composition of oxygen within the phosphate ion can improve our understanding of P cycling in soil and plant systems [24]. Analyzing the isotopic composition of oxygen bound to P ($\delta^{18}O$-P) is, however, constrained by several analytical difficulties [25].

The scientific discussion on the identification of soil organic P forms—whether soils contain simple well-identifiable organic P forms or organic P in complex macromolecular, nonidentified structures—is continuing [26]. The cycles of the biogeochemically important nutrient elements C, N, and P are closely interlinked across environmental systems.

The difference between Silicium (14) and Phosphorous (15) is just one proton, and microorganisms in soil also can dissociate the molecule of water, thereby, inside them, there are available protons and energy. Neutrons form spontaneously.

The C-, N-, P-stoichiometry of soil organic matter was primarily controlled by soil properties rather than by the elemental stoichiometry of manure or fertilizer inputs [27]. Since 1998, in Northern Germany, organic P forms in soil did not correspond with the P forms in the organic fertilizers applied to the soil [28]. It is quite possible that living organisms inside soil make the difference, given the energy, protons, and oxygen available coming from water dissociation in analogue fashion to formation of ^{14} C and ^{18} O in living beings.

Although the study of C-, N-, and P ratios is needed to understand the long-term functioning of cropped soils, it must always be tied with valuation of elemental inputs and budgets, and the capability of soils to steady, modify, and form, even a subatomic level, the C-, N-, and P-containing compounds. The soil is way beyond to be inert.

The relative importance of P fluxes arising from soil organic matter (SOM) mineralization compared with fluxes from P desorption appears to be much larger in forest and grassland than in arable soils [29], which is understandable and expected because the intensive use of agrochemicals in arable soil, and not in forest and grassland, perturbed the astonishingly accurate dissociation of the water molecule, therefore the highly ordered generation and distribution of energy, protons, and oxygen are impoverished.

Factors such as wetting and drying cycles, green manure inputs, seasonal fluctuations, amount of light, pressure, temperature, moisture, and soil parent material also clearly affect organic P mineralization [29].

The application of microbial inoculants as so-called biofertilizers has often been described as a component of sustainable nutrient management. The main efforts in this field have focused on living beings, as fungi [30].

Considering the uncertainty and the costs of microbial inoculants in practical agriculture, the activation of native soil microorganisms by agronomic measures such as organic matter management and crop rotation could be a better approach to utilize benefits of microbes [31] measures that significantly diminishing the need and use of industrialized phosphorous.

Fertilizers manufactured with high proportions of phosphorus, nitrogen, and potassium boosted plant growth to unprecedented levels, especially in tropical soils that are poor in these constituents [3]. Although phosphorus is one of the most common elements on Earth, only a small percentage is available for human use [32]. Phosphorus is seldom credited for the decline in the number of undernourished people, but such progress would have been unthinkable without its dramatically expanded use in the form of phosphate-based fertilizer.

2.2 Toxicity of phosphorous in the environment

Phosphorus plays many roles in society today—both desired and undesired [33]. Phosphorus brings about a plentiful of different functions—on immeasurably dissimilar temporal and geographical balances: transporting split-second signals to the brain in the chemical ATP, or immobile as a Ca3 (PO4)2 molecule in apatite-rich phosphate rock that took tens of millions of years to form, expecting mining, or progressively being drawn up from soil solution by plant roots via chemical dissemination, or clearing from our bodies in a momentary drop of urine before being thinned by a flood of flush water to join other domestic and industry wastewater at a distant and pestilent treatment plant, poisoning water bodies as cyanobacteria, or simply cycling naturally between land, biota, and water without being perceived by most of the society [34].

Elevated P inputs can have severe long-term effects on freshwater and marine ecosystems, and large-scale efforts are needed to reduce P inputs from land. Eutrophication is still considered to be the most serious anthropogenic threat, for instance, in the dead zone of Baltic Sea [35]. The mitigation of eutrophication in freshwater, coastal, and marine systems requires a better understanding of mobilization and release of P from soil and catchments (soil-to-water transfers), P composition and cycling in water bodies, and measures to decrease P loss.

Despite its merits as an essential staff of life, phosphorus has also a role as quite persistent pollutant. In rural areas, with poor control of agrochemicals, it often flows into receiving water (ponds, lakes, rivers, etc.) as runoff from agricultural fields and in urban areas from dirt sources as a foremost component of human body waste flushed down toilets [36]. Phosphorus can excessively boost some type of biochemical reactions meanwhile turn down others, which finally seems as endorsing and abnormal algal blooms in the lakes, ponds, and rivers where it concentrates—eutrophication [37].

The misconception that disproportionate algal growth can in the long run lower oxygen levels in the water although first the levels of dissolved oxygen fall and then it is eutrophication, it is thought that some fish species cannot tolerate this DOL of less of 6 mg/l. Thereby, the explanation of this long-lasted mistake emerges after our discovery that both prokaryotic and eukaryotic cells are phototrophs [38].

It has been demonstrated that the phosphate detergents emerging from municipal wastewater streams were a major driver of Lake Erie's problem—excessive algal growth and mortality of some fish's species [39].

Cadmium is a well-established renal, bone, and pulmonary toxicant that occurs naturally in phosphate rock deposits. Phosphate fertilizers are considered the main source of cadmium in agricultural soils [40].

In some river basins, P export now exceeds P inputs, which may result from the net mobilization of P pools accumulated during earlier decades [41], already reached a finite P-accumulation stage. Animal studies show that high inorganic phosphate feeding resulting in high serum phosphate promoted lung, skin, bladder, and prostate cancer [42].

Leaf senescence, or the final developmental stage of the leaf, means the transition from a photosynthetically active organ to the attenuation of said function and eventual death of the leaf. During senescence, essential nutrients sequestered in the leaf, such as phosphorus (P), are recycled, this is mobilized and transported to sink tissues, particularly expanding leaves and developing seeds. Phosphorus recovery is decisive, as it helps to ensure that previously acquired P is not lost to the environment, particularly under the naturally desirable occurring condition where most unfertilized soils contain low levels of soluble orthophosphate (Pi), the only form of P that roots can directly assimilate from the soil [43].

Phosphorus (P) is a key plant macronutrient, as it is a structural component of critical biomolecules involved in both temperature regulation processes, such as ATP and PPi, and in the development of key macromolecules such as nucleic acids and phospholipids. Thus, P is central to nearly all foremost metabolic processes in plants (and humans), including photosynthesis and respiration. Soluble orthophosphate (PO_4^{3-}; Pi), which is the only form of plant-available P that roots can directly assimilate from the soil, is often highly limiting in the natural environment, prompting the widespread erroneous use of Pi-containing fertilizers in agriculture [44]. And we say wrong because the living entities that live in the clay can transmute the phosphorus from the silicon, because the difference between them is just one proton.

While fertilizers are seemingly effective in boosting harvest yields, only 15–30% of applied P is on average absorbed by crops in the year of its application [45]. The resulting Pi-runoff from fertilized fields leads to nutrient overloading of aquatic ecosystems, triggering toxic algal blooms and eutrophication of the affected waterways. Furthermore, the Pi contained within these fertilizers is manufactured from nonrenewable rock-phosphate reserves, which have been projected to be depleted within the next 80 years [35].

The use of fertilizers in agricultural practices may boost efficient crop growth but could consequently inhibit efficient Pi recycling and thus the overall P-use efficiency (PUE) [46]. Nearly half of the total P present within a healthy leaf exists within nucleic acids; of that, approximately 80% is represented by ribosomal RNA (rRNA).

The possibility of microorganisms in soil can synthesize ^{15}P arising from ^{14}Si opens an unexpected source of phosphorous in minute quantities but at the same time sufficient for the fundamental biochemical needs of life, both in plants and animals. The difference between them (^{15}P and ^{14}Si) is just one proton, and water dissociation provides enough protons, energy, and high-energy electrons. As it is expected, amount of Pi resulting from this "transmutation" of elements are small, such as ^{14}C and ^{18}O, which also formed—in small quantities—inside living things, and coherently, the plants metabolism is so efficient to recycling minute amounts of Pi.

Therefore, the fertility of the soil depends significantly on its oxygen content, so much so that the formation of clays depends on the presence of oxygen and therefore on the presence of life that generates it, rather than phosphorus in irrational quantities. If we restore the oxygen levels that the arable soil should contain, soil fertility would improve significantly. And even more so if we irrigate the crops with water with dissolved oxygen levels above 6 mg / L.

Enzymes do not make possible an impossible reaction, thereby, neither water dissociation nor transmutation from ^{14}Si to ^{15}P is a biochemical reaction that depends on enzymatic activity. Thereby, Pi acquisition from the soil is not an entirely passive process due to soil's microorganisms being systems with capacities way beyond our abstraction capacity.

Zones that were found to have high heavy metal levels should be avoided to cultivate potatoes because potatoes tend to accumulate heavy metals notably higher than other types of plants. Soils that were found to be acidic traditionally should be treated with lime so that heavy metal uptake by plants via soil–plant pathway could be slowed; however, the increase of oxygen levels in the soil and the increase of pH through QBLOCK™ improve the root plant health and thereby the crops yield. It is important to protect groundwater resources in the region from heavy metal contamination especially in acidic zones [47].

Phosphate extraction increasingly generates more pollution and waste, requires more energy per nutrient value, and costs more to mine and to process [48]. The fluxes that we generate are larger than natural fluxes. This is no easy way to run a bio-geo-chemical cycle [49].

3. Discussion

First life originated in water, then glucose—the universal precursor—and thereafter phosphorus that plays important role in evolution of whole spectrum of life. Phosphorus constitutes integral part of nucleic acids and amino acids, which are carriers of the whole genetic information of evolution of life on this planet and building block in every form of life. Such is importance of element phosphorus. But now phosphorous is emerging hidden crisis before agricultural community and environmentalists of the world.

Phosphates are thermodynamically unstable while being kinetically stable [50]. ATP is kinetically stable under physiologic conditions [51]. Kornberg has estimated that eukaryotic cells contain on average 10^9 molecules of ATP [52]. Theoretically, $\approx$2.5 ATPs are formed for each pair of electrons sent by an NADH molecule down the respiratory chain and that $\approx$80 kg of ATP is turned over in a day per adult male human, thereby, about 30 kg of NADH must be generated and funneled to O_2 every day. It does not make sense.

The fluidity of nucleoside triphosphates, via NMP allocation, to handle DNA replication (~2×10^8 ATP to replicate the *E. coli* chromosome), RNA transcription (~ 200/typical mRNA of 1000 nucleotides), and the many peptide bonds in each protein (~1500 ATP per 30 kDa protein) may be the largest energy drain in proliferating cells [53]; however, ATP is not source of energy, thereby these calculations are wrong, because the source of energy of living things starts with water dissociation.

In cancer cells and in pluripotent stem cells (embryonic), TCA cycle is not fully active for purposes of ATP synthesis, even in the presence of ample oxygen. Thus, they do not oxidize glucose completely, and electrons do not get put into the mitochondrial respiratory chain effectively, the so-called Warburg effect [54].

Phosphorous is an inorganic element probably produced by photosynthesis in living beings. O_2 by analogy is an inorganic molecule also, since nearly all the 20% of the earth's atmosphere that is O_2 has been biogenically derived via O_2-producing photosynthesis.

O_2: Thermodynamically Activated, Kinetically Stable Inorganic Molecule to Power Eukaryotic Metabolism. Molecular oxygen is a difficult to handle metabolite. Living things have optimized their presence, as it has been present in the equation since the beginning of time. But the valuable product of water dissociation is hydrogen, simply because it is the quintessential energy carrier in the entire universe. Therefore, the following concepts are totally theoretical: Higher eukaryotes unlock its thermodynamic potential to undergo four-electron reduction and make a good living energetically. But glucose is not a source of energy, it is only a source of biomass, even if it is combined with oxygen. They practice substrate hydroxylation chemistry judiciously, in hypoxia, in macromolecule demethylations, and in the steroid hormone maturation pathways. Of course, oxygen appears in every reaction but its "unwanted" presence is since it comes from the dissociation of water that the body carries out to obtain energy. The oxygen that the body contains does not come from the atmosphere. Yet, they still had to evolve enzymatic and nonenzymatic defenses against toxic partially reduced oxygen metabolites, emphasizing how oxygen reductive metabolism has its intrinsic dangers. The previous paragraph, also theoretical, will have to be rewritten since the best antioxidant known is hydrogen.

The amount of ATP in a 70 kg human has been estimated at ~50 g, with about 10^9 molecules/cell.

The most common single posttranslational modification of proteins (PTM) is phosphorylation of Ser, Thr, and Tyr side chains by ATP-dependent protein kinases, but the activation energy required by enzymes comes from the dissociation of water, the effect of ATP is minor and complementary. Supposedly tens to hundreds of thousands of phosphor variants of proteins may be formed transiently in human cells by the >500 members encoded in human kinemes; however, the energy that the body obtains through the dissociation of water is exact, amazingly accurate, and has not changed since the beginning of time, so the sequence of biochemical logic with which the body handles the compounds that conform us is strictly regulated by 4 billion years of evolution. Many proteins can be phosphorylated at multiple residues, by single or multiple kinases, but it does not happen randomly, but every kinase requires energy that comes from the dissociation of the water molecule. The fraction of a given protein subject to modification can depend on location within the cell (distinct pools) available energy that comes from water dissociation and the activity of the PTM enzymes. Hundreds to thousands of fractional molecular protein variants can be created (but not randomly) and then returned to starting pools by protein phosphatases [55], which, to function properly, require adequate energy, which undoubtedly comes from the dissociation of the water molecule. By the way, any enzyme whether related to phosphate metabolism requires the energy that always comes from water dissociation. Thereby, the turnover rate of the splitting of the water molecules is the great regulator of the functioning of the biochemical logic of life both in water and in agricultural soil.

There are at least four compounds that seem to exist in abundance on planet Earth in comparison with other planets or with known Universe: oxygen, water, clay, and phosphorous.

Them all are produced by living beings. This is: the presence of these compounds and elements requires the presence of life to be produced. Without life, they are not produced or at least are absent.

Such as the O_2 present inside human body coming from water dissociation that living being has inside and not from atmosphere [56], then a significant part of phosphorous coming from the inner photosynthesis more than of diet. Phosphate esters and anhydrides dominate the living world but are seldom used as intermediates by organic chemists. Phosphoric acid is specially adapted for its role in nucleic acids because it can link two nucleotides and still ionize, something unique; the resulting negative charge serves both to stabilize the diesters against hydrolysis and to retain the molecules within a lipid membrane; but these reactions also need the power coming from water dissociation. Phosphates with multiple negative charges can react through energy expenditure, by way of the monomeric metaphosphate ion PO3- as an intermediate. No other residue appears to fulfill the multiple roles of phosphate in biochemistry; however, energy from water dissociation still is needed. Stable, negatively charged phosphates react under catalysis by enzymes—energy expenditure; organic chemists, who can only rarely use enzymatic catalysis for their reactions, need more highly reactive intermediates than phosphates.

Given our discovery of unexpected intrinsic capacity of living beings to transform sunshine power into chemical energy, through water dissociation, like plants do, we can discard the role of phosphates as energy sources [57] limiting it to temperature regulation ATP, ADP, and AMP cycle, the biology of phospho-nucleotides, and control of phosphates toxicity.

Therefore, the planetary boundary for phosphorous must be rethanked, rewritten, [58] because the abundance of phosphorous on Earth and the scarcity in Universe are not by chance. Phosphorous (^{15}P) is produced by living things, mainly by the microorganisms of the soil; probably arising from silicium (^{14}Si). Remember that the difference between them is just one proton, and the transformation of sunshine power into chemical energy is through the dissociation of the water molecules, a universal mechanism that places the adequate energy, protons, and oxygen inside every cell of living things.

4. Conclusion

Therefore, while the life thrives on planet Earth, Phosphorous should be produce by living things, as has been done since beginning of time. And it is important to respect the way nature has formed and used it, this is in minimal quantities. The secret of sustainable fertile soil lies in keeping oxygen levels high inside it; something that is possible to achieved with the QBLOCK™.

Once the knowledge about the unsuspected ability to dissociate water from living beings, to transform sunlight into chemical energy, is known and disseminated sufficiently, the use of nitrogen fertilizers can be reduced to a minimum and even stop using them completely, because the damage it causes to the environment, even from their manufacture and subsequent use, they are huge and long-lasting.

We can maintain the fertility of agricultural soil by raising the levels of oxygen it contains and irrigating crops with water with adequate levels of dissolved oxygen, this is above 6 mg/L. This would substantially reduce the need for artificial fertilizers whose synthesis alone is remarkably polluting, not to mention the amounts of phosphates that are thrown into crops and end up flowing in rivers and seas forming dead zones, as in the Gulf of Mexico, in the Gulf of Aden, the Baltic Sea, which continue to spread.

The use of QBLOCK™ or some similar method that raises the levels of dissolved oxygen in both water and agricultural soil will allow us a more rational agriculture, even regenerative, because we will be able to prevent the damage inflicted by current agrochemicals, and even reverse it.

Acknowledgements

This work was supported by Human Photosynthesis™ Study Centre. Aguascalientes 20000, México.

Author details

Arturo Solís Herrera*, María del Carmen Arias Esparza
and Martha Patricia Solís Arias
Human Photosynthesis™ Research Centre, Aguascalientes, México

*Address all correspondence to: comagua2000@yahoo.com

References

[1] Asimov I. Asimov on Chemistry. Garden City, NY: Doubleday; 1974

[2] Manz F. Why is the phosphorus content of human milk exceptionally low? Monatsschrift für Kinderheilkunde. 1992;**140**(9 Suppl. 1):S35-S39

[3] Smil V. Phosphorus in the environment: Natural flows and human interferences. Annual Review of Energy and the Environment. 2000;**25**:53-88. DOI: 10.1146/annurev.energy.25.1.53

[4] FAO. Global Hunger Declining, but Still Unacceptably High: International Hunger Targets Difficult to Reach. Rome, Italy: Food and Agriculture Organization of the United Nations; 2010. Available from: http://tinyurl.com/3bypmlv accessed 13 Apr 2011

[5] Buckingham DA, Jasinski SM. Phosphate Rock Statistics. In: Kelly TD, Matos GR, editors. Historical Statistics for Mineral and Material Commodities in the United States, Data Series 140. Washington, DC: U.S. Geological Survey. (Updated 19 October 2010). Available from: http://tinyurl.com/3ccm3ka [Accessed: 13 April 2011]

[6] According to Cordell, phosphorus lost from agricultural fields could come both from fertilizer runoff and from increased erosion of unfertilized soil containing naturally occurring phosphorus

[7] Smil V. Phosphorus: Global Transfers. Encyclopedia of Global Environmental Change. In: Douglas I, editor. Causes and Consequences of Global Environmental Change. Vol. 3. Chichester, UK: John Wiley & Sons; 2002. pp. 536-542. Available: http://tinyurl.com/5u4fy7k [Accessed: 13 April 2011]

[8] Herrera AS, Del CA, Esparza M, Md Ashraf G, Zamyatnin AA, Aliev G. Beyond mitochondria, what would be the energy source of the cell? Central Nervous System Agents in Medicinal Chemistry. 2015;**15**(1):32-41. DOI: 10.2174/1871524915666150203093656

[9] Herrera AS, Beeraka NM, Solis LFT, Mikhaleva LM, Somasundaram SG, Kirkland CE, et al. The efficacy of melanin precursor QIAPI 1© against age-related macular degeneration (AMD): A case report. Central Nervous System Agents in Medicinal Chemistry. 2020;**20**(3):218-225. DOI: 10.2174/1871524920666201109152951

[10] Herrera AS, Beeraka NM, Sinelnikov MY, Nikolenko VN, Giller DB, Solis LFT, et al. The beneficial effects of QIAPI 1® against pentavalent arsenic-induced lung toxicity: A hypothetical model for SARS CoV2-induced lung toxicity. Current Pharmaceutical Biotechnology. 2022;**23**(2):307-315. DOI: 10.2174/1389201022666210412142230

[11] Peacock M. Phosphate metabolism in health and disease. Calcified Tissue International. Jan 2021;**108**(1):3-15. DOI: 10.1007/s00223-020-00686-3. Epub 2020 Apr 7

[12] Calvo Mona S, Lamberg-Allardt CJ. Phosphorus. Advances in Nutrition (Bethesda, Md.). 2015;6:860-862. DOI: 10.3945/an.115.008516

[13] Vorland CJ, Stremke ER, Moorthi RN, Hill Gallant KM. Effects of excessive dietary phosphorus intake on bone health. Current Osteoporosis Reports. 2017;**15**(5):473-482. DOI: 10.1007/s11914-017-0398-4

[14] Cordell D, White S. Tracking phosphorus security: Indicators

of phosphorus vulnerability in the global food system. Food Security. 2015;7:337-350. DOI: 10.1146/annurevenviron-010213-113300

[15] Metson GS, MacDonald GK, Haberman D, Nesme T, Bennett EM. Feeding the corn belt: Opportunities for phosphorus recycling in US agriculture. Science of the Total Environment. 2016;**542**:1117-1126

[16] Withers PJA, van Dijk KC, Neset T-SS, Nesme T, Oenema O, Rubæk GH, et al. Stewardship to tackle global phosphorus inefficiency: The case of Europe. Ambio. 2015;**44**:193-206. DOI: 10.1007/s13280- 014-0614-8

[17] Oster M, Just F, Büsing K, Wolf P, Polley C, Vollmar B, et al. Toward improved phosphorus efficiency in monogastrics—interplay of serum, minerals, bone, and immune system after divergent dietary phosphorus supply in swine. American Journal of Physiology, Regulatory, Integrative and Comparative Physiology. 2016;**310**:917-925

[18] Von Sperber C, Tamburini F, Brunner B, Bernasconi SM, Frossard E. The oxygen isotope composition of phosphate released from phytic acid by the activity of wheat and Aspergillus niger phytase. Biogeosciences. 2015;**12**:4175-4184. DOI: 10.5194/bg-12-4175-2015

[19] Pfahler V, Dürr-Auster T, Tamburini F, Bernasconi SM, Frossard E. 18O enrichment in phosphorus pools extracted from soybean leaves. New Phytologist. 2013;**197**:186-193

[20] Herrera AS, Esparza MDCA, PES A, Ashraf GM, Mosa OF, Fisenko VP, et al. The role of melanin to dissociate oxygen from water to treat retinopathy of prematurity. Central Nervous System Agents in Medicinal Chemistry. 2019;**19**(3):215-222. DOI: 10.2174/1871524919666190702164206

[21] Italianer MF, Naninck EFG, Roelants JA, van der Horst GTJ, Reiss IKM, Goudoever JBV, et al. Circadian variation in human milk composition, a systematic review. Nutrients. 2020;**12**(8):2328. DOI: 10.3390/nu12082328

[22] Leinweber P, Bathmann U, Buczko U, Douhaire C, et al. Handling the phosphorous paradox in agriculture and natural ecosystems: Scarcity, necessity, and burden of P. Ambio. 2018;**47**(Suppl. 1):S3-S19. DOI: 10.1007/s13280-017-0968-9

[23] Rodehutscord M, Rosenfelder P. Update on phytate degradation pattern in the gastrointestinal tract of pigs and broiler chickens. In: Walk CL, Kühn I, Stein HH, Kidd MT, Rodehutscord M, editors. Phytate destruction—Consequences for precision animal nutrition. Wageningen: Wageningen Academic Publishers; 2016. pp. 15-32

[24] Tamburini F, Pfahler V, von Sperber C, Frossard E, Bernasconi SM. Oxygen Isotopes for unraveling phosphorus Ambio 2018, 47(Suppl. 1):S3–S19 www.kva.se/en 123 transformations in the soil–plant system: A review. Soil Science Society of America Journal. 2014;**78**:38-46

[25] Tamburini F, Bernasconi SM, Angert A, Weiner T, Frossard E. A method for the analysis of the d18O of inorganic phosphate in soils extracted with HCl. European Journal of Soil Science. 2010;**61**:1025-1032. DOI: 10.1111/j.1365-2389.2010.01290.x

[26] McLaren TI, Smernik RJ, McLaughlin MJ, McBeath TM, Kirby JK, Simpson RJ, et al. Complex forms of soil

organic phosphorus a major component of soil phosphorus. Environmental Science and Technology. 2015;**49**: 13238-13245

[27] Frossard E, Achat DL, Bernasconi SM, Bünemann EK, Fardeau JC, Jansa J, et al. The use of tracers to investigate phosphate cycling in soil/plant systems. In: Bünemann EK, Oberson A, Frossard E, editors. Phosphorus in action. Berlin: Springer. (Soil Biology 26); 2011

[28] Requejo M, Eichler-Löbermann B. Organic and inorganic phosphorus forms in soil as affected by long-term application of organic amendments. Nutrient Cycling in Agroecosystems. 2014;**100**:245-255

[29] Bünemann EK. Assessment of gross and net mineralization rates of soil organic phosphorus—A review. Soil Biology & Biochemistry. 2015;**89**: 82-98

[30] Vassilev N, Eichler-Löbermann B, Requena AR, Martos V, Lopez A, Vassileva M. Biodiesel by-products and P-solubilizing microorganisms. Reviews in Environmental Science and Bio/ Technology. 2016;15:627-638. DOI: 10.1007/s11157- 016-9410-1

[31] Hupfauf S, Bachmann S, Juárez MF-D, Insam H, Eichler-Löbermann B. Biogas digestates affect soil P availability and microbial community composition. Science of the Total Environment. 2016;**542**:1144-1154

[32] Schröder JJ et al. Sustainable Use of Phosphorus, Report 357. Wageningen, The Netherlands: Plant Research International, Wageningen University and Research Centre; 2010

[33] Cordell D et al. The story of phosphorus: Global food security and food for thought. Global Environmental Change. 2009;**19**(2):292-305. DOI: 10.1016/j.gloenvcha.2008.10.009

[34] Cordell D. The story of phosphorus: sustainability implications of global phosphorus scarcity for food security [doctoral thesis]. Linköping, Sweden: Linköping University; 2010. Available from: http://tinyurl.com/3fm7gjx [Accessed 13 April 2011]

[35] Nausch G, Naumann M, Umlauf L, Mohrholz V, Siegel H and Schulz-Bull D. Hydrographic-hydrochemical assessment of the Baltic Sea 2015. Meereswissenschaftliche Berichte (Warnemünde). Vol. 101. Warnemünde, Germany: Leibniz Institute for Baltic Sea Research (IOW); 2016. pp. 1-97

[36] Lougheed T. Phosphorus paradox: scarcity and overabundance of a key nutrient. Environmental Health Perspectives. 2011;**119**(5):A208-A213. DOI: 10.1289/ehp.119-a208

[37] Smil V. Phosphorus: Global transfers. In: Douglas I, editor. Encyclopedia of Global Environmental Change, Vol. 3, Causes and Consequences of Global Environmental Change. Chichester, UK: John Wiley & Sons; 2002. pp. 536-542

[38] Herrera AS, Ashraf GM, Del Carmen Arias Esparza M, Tarasov VV, Chubarev VN, Avila-Rodriguez MF, et al. Cerebrospinal fluid, brain electrolytes balance, and the unsuspected intrinsic property of melanin to dissociate the water molecule. CNS & Neurological Disorders Drug Targets. 2018;**17**(10):743-756. DOI: 10.2174/1871527317666180904093430

[39] Schindler DW. Eutrophication and recovery in Experimental Lakes: implications for lake management. Science. 1974;**184**(4139):897-899. DOI: 10.1126/science.184.4139.897

[40] Alloway BJ, editor. Heavy Metals in Soils. 2nd ed. London, UK: Blackie Academic & Professional; 1994

[41] Powers SM, Bruulsema TW, Burt TP, Chan NI, Elser JJ, Haygarth PM, et al. Long-term accumulation and transport of anthropogenic phosphorus in three river basins. Nature Geoscience. 2016;**9**:353-356

[42] Chang AR, Lazo M, Appel LJ, Gutiérrez OM, Grams ME. High dietary phosphorus intake is associated with all-cause mortality: Results from NHANES III. The American Journal of Clinical Nutrition. 2014;**99**(2):320-327

[43] Stigter KA, Plaxton WC. Molecular mechanisms of phosphorus metabolism and transport during leaf senescence. Plants (Basel). 2015;**4**(4):773-798. Published 2015 Dec 16. DOI: 10.3390/plants4040773

[44] Plaxton WC, Tran HT. Metabolic adaptations of phosphate-starved plants. Plant Physiology. 2011;**156**(3):1006-1015

[45] Baligar VC, Fageria NK, He ZL. Nutrient use efficiency in plants. Communications in Soil Science and Plant Analysis. 2001;**32**:921-950. DOI: 10.1081/CSS-100104098

[46] Veneklaas EJ, Lambers H, Bragg J, Finnegan PM, Lovelock CE, Plaxton WC, et al. Opportunities for improving phosphorus-use efficiency in crop plants. The New Phytologist. 2012;**195**(2):306-320

[47] Kara EE et al. Evaluation of heavy metals' (Cd, Cu, Ni, Pb, and Zn) distribution in sowing regions of potato fields in the province of Nigde, Turkey. Water, Air, and Soil Pollution. 2004;**153**(1-4):173-186. DOI: 10.1023/B:WATE.0000019942.37633.31

[48] Cordell D et al. Peak phosphorus: the crunch time for humanity? Sustainability Rev. 2011;**2**(2). Available from: http://tinyurl.com/69q3g7f

[49] Lougheed T. Environmental Health Perspectives. May 2011;**119**(5):585-A224

[50] Walsh CT, Tu BP, Tang Y. Eight kinetically stable but thermodynamically activated molecules that power cell metabolism. Chemical Reviews. 2018;**118**(4):1460-1494. DOI: 10.1021/acs.chemrev.7b00510. Epub 2017 Dec 22. Erratum in: Chem Rev. 2018 May 23;118(10):5261-5263. PMID: 29272116; PMCID: PMC5831524

[51] Westheimer FH. Why nature chose phosphates. Science. 1987;**235**(4793):1173-1178

[52] Kornberg A. For the Love of Enzymes: Odyssey of a Biochemist. Harvard University Press; 1989

[53] Milo R, Jorgensen P, Moran U, Weber G, Springer M. BioNumbers--the database of key numbers in molecular and cell biology. Nucleic Acids Research. 2010;**38**(Database issue):D750-3

[54] Warburg O. On the origin of cancer cells. Science. 1956;**123**(3191):309-314

[55] Shi Y. Serine/threonine phosphatases: Mechanism through structure. Cell. 2009;**139**(3):468-484

[56] Herrera AS, del Carmen M, Esparza A. Oxygen from the atmosphere cannot pass through the lung tissues and reach the bloodstream. The unexpected capacity of human body to dissociate the water molecule. Journal of Pulmonology Research & Reports. 2022;**SRC/JPRR-133**:1-4. DOI: 10.47363/JPRR/2022(4)124

[57] Herrera AS. The biological pigments in plants physiology. Agricultural Sciences. 2015:6;1262-1271. DOI: 10.4236/as.2015.610121

[58] Carpenter SR, Bennett EM. Reconsideration of the planetary boundary for phosphorus. Environmental Research Letters. 2011;**6**:014009(12pp)

Chapter 4

Alternatives to Soluble Phosphorus Fertilizers in Indian Context

Alok Singh Jayara, Rajeew Kumar, Priyanka Pandey, Manoj Kumar Bhatt, Sharad Pandey and Roshan Lal Meena

Abstract

Phosphorus is one of the primary nutrients required in crop production. Rock phosphate is the raw material required for the manufacturing of soluble phosphorus fertilizers, which is nonrenewable in nature and expected to last for 50–400 years. The restriction of resources to few geographical locations makes its supply more vulnerable. In India, 90% of the rock phosphate for fertilizer manufacturing is imported. However, the low quality of rock phosphate deposits available in India can be utilized with certain modifications in the form of addition of phosphate-solubilizing bacteria, addition of gypsum, and in the form of phospho-enriched compost. Agriculture, livestock, urban and industrial waste can also prove to be a source of phosphorus through crystallization of struvite. There are encouraging results of struvite compared with soluble phosphorus fertilizers. This will reduce the import dependency in India as well as encourage the *Atmanirbhar* initiative in phosphorus fertilizer.

Keywords: phosphorus, nonrenewable, rock phosphate, phosphate-solubilizing bacteria, gypsum, phsopho-enriched compost, struvite

1. Introduction

The term phosphorus derived from the Greek word "Phos" meaning light, and "phorus" means bearer. Elemental form of phosphorus was discovered by German Alchemist, Henning Brandt in 1669 [1, 2]. Phosphorus evolved from seventeenth century as philosopher's stone to medicinal phosphorus, flammable phosphorus, essential nutrient in crop production, element of war, cause of eutrophication to its emerging scarcity in recent twenty-first century [2]. Earlier it had been established that adding ground bone increases the crop yield and subsequently Lawes (1842) patented the process of phosphate solublization [3]. With the proposition of Criteria of Essentiality (1939) by Arnon and Stout, it had been established that the roles and functions of each essential nutrient are irreplaceable in plant system. Phosphorus is one of the essential nutrients for plant, which extends to animals also [2]. Around 80% of the world phosphorus is utilized in agriculture [1]. Rock phosphate is one of the basic raw materials for the synthesis of phosphatic fertilizers [4]; however, its nonrenewable nature increases the vulnerability in the long term.

Phosphorus in Indian soils occurs primarily in inorganic form contributing 54–84% of the total phosphorus and organic contributing 16–46% [5]. More than 90% districts in India are classified under low-to-moderate phosphorus availability [5, 6]. The phosphorus status of soil doesn't necessarily reflects its availability to the crop plants, which is governed by the presence of calcium, iron, and aluminum phosphates; thus, only 30% of the soil phosphorus is utilized by the crop and rest remains in the soil [7]. Therefore, it demands the external application of phosphorus through fertilizers. India imports the high-grade rock phosphate for the synthesis of soluble phosphorus fertilizers. This import cost along with the decontrol of the phosphorus and potassium fertilizers in 1992 has led to the fertilizer application in favor of urea. The turmoil in the Former Soviet region in present times has also increased the vulnerability of the rock phosphate import in near future. Therefore, it is important to increase self-reliance in the field of phosphorus fertilizer application. It has been reported that indigenous rock phosphate if suitable is treated with solubilizing microorganisms or acidulates; there can be increased solublization and availability of phosphorus to the plants and consequent yields [8, 9]. Precipitation of struvite from the agriculture and livestock waste can prove another efficient alternative [10]. The major objectives of the following discussion in chapter are to find out in ways and means to enhance the dependability of phosphorus fertilizer on indigenous resources in general and on farm resources in particular.

2. Importance of phosphorus in crop production

Phosphorus constitutes 0.2% (0.1–0.5%) in the plant system. Most of the phosphorus is absorbed in the form of primary orthophosphate ions; however, it is also absorbed in the form of secondary orthophosphate. Phosphorus is not reduced like nitrates and sulfates and rather exists as inorganic phosphate or esterifies to carbon chain through hydroxyl group or attaches to another phosphate group through energy-rich pyrophosphate bonds [11].

The major functions of phosphorus are as structural element of nucleic acids; phospholipids of biomembrane forming bridge between triglyceride and other molecules; energy-rich phosphates and phosphate esters in metabolism; acts as regulator in glycolysis, photosynthesis, respiration, nitrogen assimilation, starch synthesis in chloroplast; detoxification of heavy metals by binding with phytates [11, 12]. Phosphorus control over photosynthesis involves ratio of Pi to triose phosphate; light activation of ribulose bisphosphate carboxylase; activation of fructose-1,6-bisphosphatase, sedoheptulose 1,7-bisphosphatase; ATP/ADP ratio; decreased regeneration of RuBP, and low sink strength under P-deficient conditions leading to reduced photosynthesis [12]. The stored inorganic phosphate in the plant varies according to its availability; however, phosphorus in metabolism remains stable with former acting as buffer [13]. Phosphorus supply is more critical in the early season as observed in various annual crops; however, later stage supplementation improves yield though plant is also able to remobilize the stored phosphates to the grains [13].

The similar role can be played by higher seed phosphorus content where it will supply early seedling growth P requirement leading to better root development and thus giving access to growth limiting water and mineral nutrients [14]. Phosphorus is required more for the nodule growth and nitrogenase activity in the N-fixing plants than for the whole plant growth [15, 16]. Phosphorus supply increases the root diameter and dry weight; however, increased root shoot diameter, root hair length and density, root branching, root hair are observed under P-deficient conditions to increase P acquisition [17, 18].

3. Need for the alternatives in Indian context

Before the utilization of phosphate rocks for commercial fertilizer production, bones, corpolites (fossilized animal dung), and guano were the main sources of phosphorus supplementation [3]. It is evident that various sources of phosphorus had been utilized including crop residues; human, animal, fish, and bird waste from Middle East to Japan [2].

The process of super phosphate production by solubilizing bones in sulfuric acid was separately patented by J B Lawes and James Murray in 1842 leading to the development of superphosphate and mixed fertilizer industry [3, 19]. Phosphorus consumption in the post-World War II era was not that much intensified; however, with introduction of Green Revolution after mid-twentieth century, there was increase in use of phosphorus fertilizer along with nitrogen and potassium owing to introduction of the fertilizer-responsive varieties. Both annual phosphate rock extraction and per capita production have seen consistent growth of 3–4% and 1.4%, respectively, with an increase of more than 300% phosphorus fertilizer consumption between 1961 and 2013, however, characterized by a decline post 1989 for a considerable period owing to disintegration of Soviet Union and decreased fertilizer demand in Western Europe and North America [19, 20]. More than 70% of the world phosphorus reserves are located in South Africa, Morocco and Western Sahara and United States [21] and Brazil and Peru in South America; China, Iraq, Israel, and Jordan in Asia; Australia in Oceania; Former Soviet Union in Europe are some major countries continent-wise [22]. The mineral resource extraction follows a mountain/ U-shaped curve where there is initial increase followed by the plateau and then decline in production with time [22]. Phosphorus reserve exploitation has been correlated with time, and it is considered to have "peak P" analogous to the "peak oil" by 2035, after which demand will outstrip supply [3, 23]. With current utilization rate, the phosphate rock reserves are expected to last for 50–400 years [24, 25].

In India, 90% of the rock phosphate required for soluble P fertilizer manufacturing is imported, out of which 80% is imported from Jordan, Morocco, and Egypt [26]. The nonrenewable nature along with its restricted availability to few countries in the world increases the vulnerability of its supplies in case of any untoward incident happens in these nations. The present fertilizer use is also characterized by the imbalance with $N:P_2O_5: K_2O$ use ratio for 2018–19 as 7.1:2.7:1 [27] in place of recommended ratio of 4:2:1. It can be partly attributed to the higher cost of import of phosphorus and potassic fertilizers. Fertilizer subsidy has seen increase of more than 200% since 2010–11 when it reached 1.34 lakh crores in 2020–21 [28]. This becomes more relevant in present times with the turmoil arising in Former Soviet Union region, which is dwindling the resources supply, and the rising protectionism approach of the nations. Therefore, it is important to seek alternate sources of P fertilization in India, which are more indigenous in nature. In addition to reducing the cost of farming, the concept will also support the *Atmanirbhar Bharat* and self-reliance as it will be more depending on the indigenous minerals and on-farm waste generated in the farmers' field.

4. Various alternatives of phosphatic fertilizers

4.1 Indigenous rock phosphates

India has substantial reserves of low-grade phosphorus [29, 30], which can be utilized as nutrient source. Total resource of rock phosphate in India is estimated at 312.67 mT of which only 45.80 mT constitutes reserve [31]. Of the total resources, only 8% constitutes fertilizer grade and around 37% is classified under low-grade reserve [31]. However, direct application of rock phosphate is only suited for acidic soils, not for the neutral to alkaline soils where pH is more than 5.5–6.0, and is less reactive, of low grade, and has poor agronomic efficiency [32–36].

The acidulation of these rock phosphates either with the sulfur-based minerals or sulfur or organic matter holds promise to increase its availability in the soil. Addition of rock phosphate to compost or straw can be useful in increasing the phosphorus availability to the crop. The composting process leads to the mobilization of phosphorus in the rock phosphate, in addition to supplying secondary and micronutrients [37].

Partial acidulation of rock phosphate with locally sourced material such as farm residues, manure, and compost is cost-effective technique to make phosphorus available by the release of chelating action forming complexes and humic acid [38].

4.1.1 Addition of phosphate-solubilizing bacteria (PSBs)

An important aspect of Rock phosphate solublization is addition of phosphate-solubilizing microorganisms where they release organic acids, chelation, and increase in phosphatase enzyme activity and ion exchange reactions and thus, increase the availability [35, 36, 38]. There are various species of phosphate-solubilizing bacteria including *Serratia, Burkholderia, Azotobacter, Rhodococcus, Xanthomonas, Enterobacter,* and *Pantoea* [39]. Rock phosphate when incubated with PSB has shown encouraging results (**Table 1**).

4.1.2 Phosphate-enriched compost/manure

Composting is common practice in India; however, it is characterized with lower nutrient concentration. Around 500 mT of crop residues is generated in India of which three-fourth is contributed by the cereals. Enrichment of compost with rock phosphate has twin benefits with enriched nutrient content and comparatively more solublization of phosphate mineral. Addition of rock phosphate is found to not only increase the nutrient content of rice straw compost but also reduce C:N ratio when added in combination of waste mica [42].

Addition of rock phosphate is reported to reduce total C content in composting mass due to dilution effect, increase in total nitrogen content due to net loss of dry matter, and significant increase in phosphate content with higher share of citrate soluble form than water-soluble as it contributed by the rock phosphate and favorable maturity index such as C/N ratio < 20 except for tree leaf compost due to higher cellulose, hemicelluloses, and lignin content; water-soluble carbon to organic nitrogen ratio < 0.5; nitrification index> 0.16 [43–45].

Higher reduction in C:N and C:P ratio and increase in water-soluble P were recorded in rock-phosphorus-enriched manure when inoculated with phosphate-solubilizing microorganisms [34] with citric, malic, oxalic, and formic acids with citric acid having maximum P-solubilizing efficiency [46]. Higher rate of nitrogen

Sr. No.	Crop	Modification of rock phosphate	Observations	References
1.	Potato-soybean	Rice straw enriched with Rock phosphate, mica and *Aspergillus awamori* (along with 50 and 75% of RDF)	Significantly higher yield over RDF and control up to 43.3 and 21.5%, respectively in potato and 27.6–46.9% increase in soybean grain yield.	[37]
2.	Maize-Wheat	Rock phosphate treated with *Pantoea cypripedii* (PSB-3) and *Pseudomonas plecoglossicida* (PSB-5)	Significantly higher yield for both inoculants in maize compared with DAP application and for wheat though highest with RP + inoculants, however, at par with DAP	[35]
3.	Maize-Wheat	Rock phosphate treated with *Penicillium oxalicum*	Significantly higher yield for both maize and wheat compared with control and rock phosphate alone.	[33]
4.	Soybean-Wheat	Half of recommended P by rock phosphate with *Pseudomonas striata* and *Glomus fasciculatum*	Significantly higher yield compared to control, highest in soybean (3.4% higher than when 100% is through SSP); and almost similar in wheat.	[40]
5.	Rice-rapeseed-mungbean	Mussorie rock phosphate (MRP) inoculated with *Pseudomonas striata*	In 3 year experiment, MRP @ 17.5 kg P ha^{-1} was found significantly superior to control in all 3 years and at par with DAP @ 17.5 kg P ha^{-1} in third year in terms of rice equivalent yields.	[29]
6.	*Stevia rebaudiana*	Mussorie rock phsophate (MRP) treated with PSB (*Burkholderia gladioli; Enterobacter aerogenes* and *Serratia marcescens*)	Increase of 136% total biomass compared with absolute control and stevioside and rebaudioside-A recorded 291 and 575% increase, respectively.	[41]

Table 1.
Research evidences showing utilization of rock phosphate with PSB strain in different crops. The comparison for the crops/cropping system is done with control and commercial phosphorus fertilizer, and summarized findings are mentioned.

and sulfur mineralization has been observed with rock-phosphate-enriched compost along with 50% NPK after 120 days and significant improvement in available P with progress of incubation period indicating its availability for longer period of time compared with 100% NPK [45, 47].

In cropping systems, residual effect of enriched compost is found to significantly increase the total phosphorus, enhance grain and stover yield and P uptake and its use efficiency compared with similar P dose from phosphorus fertilizer [45, 48]. Rock-phosphate-enriched compost reported significantly highest grain yield and P uptake in cowpea when compared with other treatments including P fertilizer as sole source [49].

Application of rock-phosphate-enriched compost along with RDF sharing equal to the phosphorus dose in rice recorded significantly higher labile P and grain yield, only at par with treatment having total phosphorus dose from RDF [49]. Phosphorus-enriched manure has significant effect on the number of nodules, their fresh and dry weight in legume crop and yields at par with soluble fertilizer when these are combined with soluble fertilizer in equal proportion [50, 51]. Among all the phosphorus-enriched manures, highest pod and stover yield of mung bean has been reported with that inoculated with *Penicillium oxalicum* [34]. Phosphate-rich organic manure (PROM) and rock phosphate mixed with urea have shown the comparative results with soluble phosphates and even higher yield under residual effect showing the long-term availability of phosphorus [52]. The effect of phospho-enriched compost on various crops is presented in **Table 2.**

4.1.3 Acidulation with gypsum

Rock phosphate is made to react with sulfuric acid for production of single super phosphate. Similarly, gypsum can also serve as source of partial acidulation of rock phosphate. Total reserve of gypsum in India stands at 36.6 mT with 80% of the total reserve under fertilizer grade and Rajasthan, Jammu & Kashmir, Tamil Nadu, and Gujarat as the major states [31]. Low-grade gypsum is utilized as soil amendment for the sodic soils.

It can also be utilized as an acidulate for the rock phosphate treatment. In addition to the extraction of phosphorus from rock phosphates, gypsum serves as source of

Sr. No.	Crop	Particular of PEC	Observations	References
1.	Blackgram	PEC prepared with rock phosphate, mica, maize straw and FYM along with PSB inoculation.	Significant seed yield recorded upto 4 t ha^{-1} with 15.21% increase over control and at par with 6 t ha^{-1}.	[53]
2.	Cowpea	Rock phosphate was added to biowaste according to P_2O_5 requirement of cowpea	83% yield gain over control and 55% gain over soluble P fertilizer	[54]
3.	Chickpea and lentil	Rock phosphate was added to composting material consisting of fruit peels and vegetable waste	Nodules $plant^{-1}$, fresh and dry weight of nodules was higher and 15% higher yields compared to control (P fertilizer).	[51]
4.	Wheat-Greengram rotation	Rock phosphate was added to rice residue, mustard leaf residue and tree leaf separately	No significant difference with different composts on grain yields for both crops. Enriched compost of rice residue reported highest yields and residual effect on successive greengram.	[44]
5.	Rice	Rock phosphate was added to rice straw and aerobic composting was done.	Enriched compost along with chemical fertilizer (50:50) reported numerically higher yield compared when P was supplied through fertilizer alone.	[49]

Table 2.
Effect of phospho-enriched compost on various crops/cropping system. Source of organic matter in compost is indicated and yields as well as nodule growth are mentioned in comparison to control/soluble phosphatic fertilizers.

calcium and sulfur to the crop plants. Among the various acidulates tested with rock phosphate in various *Kharif* and *Rabi* crops, gypsum acidulated rock phosphate has higher yields in most of crops with significant results in Indian mustard [55].

Acidulation of rock phosphate with gypsum in wheat crop reported highest though nonsignificant among all acidulates and more than the treatment having soluble P fertilizer source [56]. Gypsum along with PSB had significantly higher yield sweet pepper yield when applied with rock phosphate compared with its use without gypsum [57].

4.2 Precipitation of struvite mineral

Struvite, a crystalline mineral having formula $MgNH_4PO_4.6H_2O$, has equimolar concentration of phosphate, ammonium, and magnesium ions [58, 59] and thus can prove as an alternate source of phosphorus. Struvite can be crystallized from the wastewaters of agriculture, sewage effluents, industrial streams, animal waste, and urine [58]. It has various properties, which make it suitable for use in agriculture (**Table 3**).

Around 85% of the mined phosphorus finds way in the water bodies through soil erosion, agriculture, and livestock waste, which promotes eutrophication; therefore, struvite precipitation can be useful to reduce this pollution [61]. Two major conditions for struvite crystallization are pH between 8.5 and 9.5 with maximum at 9.5, which reduces further, and the concentration of three ions in equimolar concentration and above struvite solubility limit ($>0.2\ gl^{-1}$) [58–62]. P content of the struvite remains in the range of 11–26% depending on the source, of which 1–2% is water-soluble and slow release in nature [60]. According to an estimate, taking stock of cow urine generated in India, 12000 tonnes of struvite can be produced daily and enlarged scope when considered for all the livestock population [61]. Similar potential exists for the industrial and domestic waste water in India. Globally, full-scale struvite recovery plants are functional in countries of Europe, North America, and Japan [61].

The extraction of struvite has been from the various sources. It is now utilized as the fertilizers or mixed with other fertilizers for value addition having good market and used in crops such as paddy, vegetables, and flowers and even reported to increase quality of paddy [63]. Performance of crops when applied with struvite when compared with chemical fertilizers is presented in **Table 4**.

Sr. No.	Parameter	Struvite characteristics
1.	Nutrient content	Rich in nitrogen, phosphorus, and magnesium
2.	Pattern of release	Slow release
3.	Influence of pH	Not soluble in alkaline soil and effective in neutral and acidic soils
4.	Suitability	Suitable for crop requiring high Mg and P doses
5.	Comparative advantages	Significantly higher dissolution rate compared with other P minerals such as fluorapatite and variscite due to weak H bonds; N leaching rates are significantly lower when compared with other N fertilizers

Table 3.
The major characteristics of struvite as fertilizer in relation to crop production [60].

Sr. No.	Source	Control fertilizer	Crop	Remarks	Reference
1.	Distillery waste water	DAP	*Brassica alba*	15% increase in dry weight, 3.2% increase in P uptake, 49% increase in chlorophyll content.	[64]
2.	Commercial	Mineral (KPO_4H_2)	*Phaseolus vulgaris*	Application of struvite more than 5 g plant^{-1} leads to higher yield than mineral fertilizer.	[59]
3.	Human urine	DAP	*Arachis hypogea*	Seed yield and other yield attributing characteristics were significantly higher with DAP inoculated with Nitrogen fixing bacteria, however, similar inoculation of struvite reported significantly higher parameters than DAP alone.	[62]
4.	Livestock wastewater	Urea and Magnesium sulfate	*Vegetable crops in pot*	It was superior to control fertilizer for all the vegetables. The increase in dosage was less inhibitory than control fertilizer.	[65]
5.	Manure	Mono Ammonium Phosphate (MAP)	*Canola*	At lower rates, the mean biomass yields were similar to control which decreased at higher rates; Quadratic response for P uptake for struvite with increase in dose whereas it was linear with MAP.	[66]
6.	Commercial	Struvite and MAP were mixed in different ratios in gradation.	*Maize-Soybean*	Upto 50% struvite blending reported statistically similar biomass which decreased further and lowest with 100% struvite in maize and upto 25% for soybean, however, it was 50% for only root biomass for it.	[67]
7.	Cow urine	DAP	*Vigna radiata*	At higher rates struvite application reported significant increase in the leaf area, stem and root dry weight and total chlorophyll content.	[58]

Table 4.
Performance of struvite as phosphorus source in various crops. Source of extraction is mentioned and commercial fertilizers are used as control. The effect of struvite on chlorophyll is also mentioned as it is source of magnesium.

4.3 Steel slag

Steel slag is the by-product of steel industry. In total, 150–180 kg steel slag is generated for 1 ton of steel production in India [68] and has been estimated at 39 mT for 2017–18 [69]. Per capita steel consumption is expected to increase more than double between 2018 and 2030–31. Thus, there exists a huge potential of generation of steel slag. In countries such as United States, Japan, and European countries, more than

80% of steel slag is recycled, whereas for countries such as China and India, it is less than 30% [70]. While slag is rich source of calcium, silicon, iron, it also reported to contain up to 4% phosphorus [71]. Application of slag is reported to increase the dry weight, yield, and phosphorus uptake in maize [72]. In addition to this, application of steel slag is advantageous in increasing the quality of the produce, source of silica to the crops, promotes immobilization of heavy metals, reduction of disease incidence in crops, and promoting carbon sequestration and reducing methane emission and can act as potential liming material [73–78].

5. Conclusion

Phosphorus is a valuable element in relation to agriculture; however, its depleting reserve presents a potential challenge to nations across the world. India is at much vulnerable position as there is huge import dependency for its raw material. Keeping the situation in view, Government of India has introduced the action plan to make country *Atmanirbhar* in phosphorus fertilizer production by utilizing the existing rock phosphate indigenous resources in country. It can be concluded that with certain modifications, indigenous low-quality rock phosphate can also be enriched and the phosphorus can be made available to the crops. The utilization of struvite is still need to be explored in India. Though there is commercial exploitation of this as phosphorus source in countries across the world, there is need to have extensive work on its extraction and develop low cost and customized approach. This will not only reduce the import cost but will also promote the sustainability at village level and concept of circular economy, which is expected to bring benefit of 40 lakh crores in 2050.

Author details

Alok Singh Jayara[1*], Rajeew Kumar[1], Priyanka Pandey[2], Manoj Kumar Bhatt[1], Sharad Pandey[3] and Roshan Lal Meena[4]

1 Department of Agronomy, G.B. Pant University of Agriculture and Technology, Pantnagar, Uttarakhand, India

2 Department of Molecular Biology and Genetic Engineering, G.B. Pant University of Agriculture and Technology, Pantnagar, Uttarakhand, India

3 School of Agriculture, Himgiri Zee University, Dehradun, Uttarakhand, India

4 National Bureau of Soil Science and Land Use Planning, Udaipur, Rajasthan, India

*Address all correspondence to: aloksingh.jayara@gmail.com

References

[1] Gleason W. An introduction to phosphorus: History, production, and application. JOM. 2007;**59**(6):17-19

[2] Ashley K, Cordell D, Mavinic D. A brief history of phosphorus: From the philosopher's stone to nutrient recovery and reuse. Chemosphere. 2011;**84**(6):737-746

[3] Ulrich AE, Frossard E. On the history of a reoccurring concept: Phosphorus scarcity. Science of the Total Environment. 2014;**490**:694-707

[4] Samreen S, Kausar S. Phosphorus Fertilizaton: The original and commercial sources. In: Tao Zhang T, editor. Phosphorus—Recovery and Recycling. London, UK: Intechopen Limited; 2018

[5] Sanyal SK, Dwivedi BS, Singh VK, Majumdar K, Datta SC, Pattanayak SK, et al. Phosphorus in relation to dominant cropping sequences in India: Chemistry, fertility relations and management options. Current Science. 2015;**2015**:1262-1270

[6] Dey P, Santhi R, Maragatham S, Sellamuthu KM. Status of phosphorus and potassium in the Indian soils vis-à-vis world soils. Indian Journal of Fertilizers. 2017;**13**(4):44-59

[7] Shah TI, Rai AP, Ma A. Relationship of phosphorus fractions with soil properties in mothbean growing acid soils of north western Indian himalayas. Communications in Soil Science and Plant Analysis, 2019. 2019;**50**(9):1192-1198

[8] Biswas DR. Nutrient recycling potential of rock phosphate and waste mica enriched compost on crop productivity and changes in soil fertility under potato–soybean cropping sequence in an Inceptisol of Indo-Gangetic Plains of India. Nutrient Cycling in Agroecosystems. 2011;**89**(1):15-30

[9] Beura K, Ghosh GK, Pradhan AK, Kohli A. Forms of phosphorus and its bioavailability in rice grown in an alluvial soil treated with rock phosphate enriched compost. Journal of Plant Nutrition. 2021;**2021**:1-21

[10] Vasa TN, Chacko SP. Recovery of struvite from wastewaters as an eco-friendly fertilizer: Review of the art and perspective for a sustainable agriculture practice in India. Sustainable Energy Technologies and Assessments. 2021;**48**:101573

[11] Hawkesford M, Horst W, Kichey T, Lambers H, Schjoerring J, Skrumsager M, et al. Functions of macronutreint. In: Marschner H, editor. Mineral Nutrition of Higher Plants. 3rd ed. London, UK: Academic Press; 2012. pp. 135-189

[12] Rychter AM, Rao IM. Role of phosphorus in photosynthetic carbon metabolism. In: Handbook of Photosynthesis. Vol. 2. 2005. pp. 123-148

[13] Grant CA, Flaten DN, Tomasiewicz DJ, Sheppard SC. The importance of early season phosphorus nutrition. Canadian Journal of Plant Science. 2011;**81**(2):211-224

[14] White PJ, Veneklaas EJ. Nature and nurture: The importance of seed phosphorus content. Plant and soil. 2012;**357**(1):1-8

[15] Jakobsen I. The role of phosphorus in nitrogen fixation by young pea plants (Pisum sativum). Physiologia Plantarum. 1985;**64**(2):190-196

[16] Israel DW. Investigation of the role of phosphorus in symbiotic dinitrogen fixation. Plant Physiology. 1987;**84**(3):835-840

[17] Shen J, Yuan L, Zhang J, Li H, Bai Z, Chen X, et al. Phosphorus dynamics: From soil to plant. Plant Physiology. 2011;**156**(3):997-1005

[18] Heydari MM, Brook RM, Jones DL. The role of phosphorus sources on root diameter, root length and root dry matter of barley (Hordeum vulgare L.). Journal of Plant Nutrition. 2019;**42**(1):1-15

[19] Stewart WM, Hammond LL, Van Kauwenbergh SJ. Phosphorus as a natural resource. Phosphorus: Agriculture and the Environment. 2005;**46**:1-22

[20] Chen M, Graedel TE. A half-century of global phosphorus flows, stocks, production, consumption, recycling, and environmental impacts. Global Environmental Change. 2016;**36**:139-152

[21] Kirkby EA, Johnston AE. Soil and fertilizer phosphorus in relation to crop nutrition. In: The Ecophysiology of Plant-phosphorus Interactions. 2008. pp. 177-223

[22] Wellmer FW, Scholz RW. Peak minerals: What can we learn from the history of mineral economics and the cases of gold and phosphorus? Mineral Economics. 2017;**30**(2):73-93

[23] Cordell D, Rosemarin A, Schröder JJ, Smit AL. Towards global phosphorus security: A systems framework for phosphorus recovery and reuse options. Chemosphere. 2011;**84**(6):747-758

[24] Schroder JJ, Smit AL, Cordell D, Rosemarin A. Improved Phosphorus use efficiency in agriculture: A key requirement for its sustainable use. Chemosphere. 2011;**84**(6):822-831

[25] Dhillon J, Torres G, Driver E, Figueiredo B, Raun WR. World Phosphorus use efficiency in cereal crops. Agronomy Journal. 2017;**109**(4):1670-1677

[26] IBM. Apatite and Rock Phosphates. In: Indian Mineral Yearbook. 58th ed. Ministry of Mines, Govt of India; 2019

[27] FAI. Annual Review of Fertilizer Production and Consumption, Executive Summary. 2019. 1049-1051

[28] FAI. Central Subsidy on Fertilizers, (2010-11 to 2021-22). 2022

[29] Sharma SN, Prasad R, Shivay YS, Dwivedi MK, Kumar S, Davari MR, et al. Relative efficiency of diammonium phosphate and mussoorie rock phosphate on productivity and phosphorus balance in a rice–rapeseed–mungbean cropping system. Nutrient Cycling in Agroecosystems. 2010;**86**(2):199-209

[30] Shivay YS. Effect of diammonium phosphate and mussoorie rock phosphate on productivity and economics of potato (Solanum tuberosum). Indian Journal of Agricultural Sciences. 2010;**80**(4):329-332

[31] IBM. Minor Minerals 30.12 Gypsum (minor) and Selenite (Major). In: Indian Mineral Yearbook. 57th ed. Ministry of Mines, Govt of India; 2018

[32] Biswas DR, Narayanasamy G. Rock phosphate enriched compost: An approach to improve low-grade Indian rock phosphate. Bioresource Technology. 2006;**97**(18):2243-2251

[33] Singh H, Reddy MS. Effect of inoculation with phosphate solubilizing fungus on growth and nutrient uptake of wheat and maize plants fertilized with rock phosphate in alkaline soils. European Journal of Soil Biology. 2011;**47**(1):30-34

[34] Shrivastava M, Kale SP, D'Souza SF. Rock phosphate enriched post-methanation bio-sludge from kitchen waste based biogas plant as P source for mungbean and its effect on rhizosphere phosphatase activity. European Journal of Soil Biology. 2011;**47**(3):205-212

[35] Kaur G, Reddy MS. Effects of phosphate-solubilizing bacteria, rock phosphate and chemical fertilizers on maize-wheat cropping cycle and economics. Pedosphere. 2015;**25**(3):428-437

[36] Yadav H, Fatima R, Sharma A, Mathur S. Enhancement of applicability of rock phosphate in alkaline soils by organic compost. Applied Soil Ecology. 2017;**113**:80-85

[37] Cicek H, Bhullar GS, Mandloi LS, Andres C, Riar AS. Partial acidulation of rock phosphate for increased productivity in organic and smallholder farming. Sustainability. 2020;**12**(2):607

[38] Ghosh R, Mandal NC. Use of plant growth-promoting Burkholderia species with rock phosphate–solubilizing potential toward crop improvement. In: Microbial Services in Restoration Ecology. Elsevier; 2020. pp. 139-156

[39] Mahanta D, Rai RK, Dhar S, Varghese E, Raja A, Purakayastha TJ. Modification of root properties with phosphate solubilizing bacteria and arbuscular mycorrhiza to reduce rock phosphate application in soybean-wheat cropping system. Ecological Engineering. 2018;**111**:31-43

[40] Gupta M, Bisht S, Singh B, Gulati A, Tewari R. Enhanced biomass and steviol glycosides in Stevia rebaudiana treated with phosphate-solubilizing bacteria and rock phosphate. Plant Growth Regulation. 2011;**65**(3):449-457

[41] Biswas DR, Narayanasamy G, Datta SC, Singh G, Begum M, Maiti D, et al. Changes in nutrient status during preparation of enriched organomineral fertilizers using rice straw, low-grade rock phosphate, waste mica, and phosphate solubilizing microorganism. Communications in Soil Science and Plant Analysis. 2009;**40**(13-14):2285-2307

[42] Moharana PC, Biswas DR. Assessment of maturity indices of rock phosphate enriched composts using variable crop residues. Bioresource Technology. 2016;**222**:1-13

[43] Moharana PC, Biswas DR. Use of mineralization kinetics to estimate the potentially mineralizable nitrogen of rock phosphate-enriched composts-amended soil. Journal of Plant Nutrition. 2018;**41**(10):1333-1344

[44] Moharana PC, Biswas DR. Phosphorus delivery potential in soil amended with rock phosphate enriched composts of variable crop residues under wheat–green gram cropping sequence. Communications in Soil Science and Plant Analysis. 2022;**2022**:1-18

[45] Kumari A, Kapoor KK, Kundu BS, Kumari MR. Identification of organic acids produced during rice straw decomposition and their role in rock phosphate solubilization. Plant Soil and Environment. 2008;**54**(2):72

[46] Moharana PC, Biswas DR, Datta SC. Mineralization of nitrogen, phosphorus and sulphur in soil as influenced by rock phosphate enriched compost and chemical fertilizers. Journal of the Indian Society of Soil Science. 2015;**63**(3):283-293

[47] Reza SK, Pal S, Singh S. Rock phosphate-enriched pressmud compost: Direct effect in pearl millet (Pennisetum

glaucum L.) and residual effect in mustard (Brassica juncea) in a Typic Haplustept. 2012

[48] Kumar A, Singhal SK, Singh V, Kumar N, Sharma VK. Impact of rock-phosphate enriched pressmud and biogass swrry on yield, phosphorus nutrition and utilization by soybean (Glycine max) in a typic hapwstept. Legume Research-An International Journal. 2013;**36**(1):79-83

[49] Ditta A, Imtiaz M, Mehmood S, Rizwan MS, Mubeen F, Aziz O, et al. Rock phosphate-enriched organic fertilizer with phosphate-solubilizing microorganisms improves nodulation, growth, and yield of legumes. Communications in Soil Science and Plant Analysis. 2018;**49**(21):2715-2725

[50] Sekhar DMR, Katewa MK, Shaktawat MS. PROM Khad an efficient source of P to replace the costly chemical Phosphatic Fertilizer. Nature Proceedings. 2012:1-1

[51] Meena M, Jat G, Meena RH, Choudhary R, Jain D, Doodhwal K, et al. Effect of phospho enriched compost and zinc on productivity and nutrient uptake of blackgram (Vigna mungo L.) in Sub-humid Southern Hills and Aravalli Region of Rajasthan. Legume Research. 2021;**45**(2):203-208. DOI: 10.18805/LR-4359

[52] Kumari MS, Ushakumari K. Effect of vermicompost enriched with rock phosphate on the yield and uptake of nutrients in cowpea (*Vigna unguiculata* [L.] Walp). Journal of Tropical Agriculture. 2006;**40**:27-30

[53] Kumar R, Krishna M, Bhatnagar A, Pandey DS, Singh VP, Ram P. Effect of rock phosphate with different acidulates on growth and yield of different field crops. Indian Journal of Agronomy. 2018;**63**(3):293-299

[54] Chaudhary SK, Kumar R, Singh AK, Kumar R. Effect of acidulated rock phosphate on growth yield attributes and yield of wheat (*Triticum aestivum* L.). Indian. Journal of Agricultural Research. 2015;**49**(6):574-576

[55] Kamal AM. Effect of gypsum, phosphoreine and rock phosphate on growth and yield of sweet pepper plants. Journal of Plant Production. 2008;**33**(12):8709-8722

[56] Prabhu M, Mutnuri S. Cow urine as a potential source for struvite production. International Journal of Recycling of Organic Waste in Agriculture. 2014;**3**(1):1-12

[57] Arcas-Pilz V, Rufí-Salís M, Parada F, Petit-Boix A, Gabarrell X, Villalba G. Recovered phosphorus for a more resilient urban agriculture: Assessment of the fertilizer potential of struvite in hydroponics. Science of the Total Environment. 2021;**799**:149424

[58] Kataki S, West H, Clarke M, Baruah DC. Phosphorus recovery as struvite: Recent concerns for use of seed, alternative Mg source, nitrogen conservation and fertilizer potential. Resources, Conservation and Recycling. 2016;**107**:142-156

[59] Sathiasivan K, Ramaswamy J, Rajesh M. Struvite recovery from human urine in inverse fluidized bed reactor and evaluation of its fertilizing potential on the growth of Arachis hypogaea. Journal of Environmental Chemical Engineering. 2021;**9**(1):104965

[60] Kumari S, Jose S, Tyagi M, Jagadevan S. A holistic and sustainable approach for recovery of phosphorus via struvite crystallization from synthetic distillery wastewater. Journal of Cleaner Production. 2020;**254**:120037

[61] Ucno Y, Fujii M. Three years experience of operating and selling recovered struvite from full-scale plant. Environmental Technology. 2001;**22**(11):1373-1381

[62] Min KJ, Kim D, Lee J, Lee K, Park KY. Characteristics of vegetable crop cultivation and nutrient releasing with struvite as a slow-release fertilizer. Environmental Science and Pollution Research. 2019;**26**(33):34332-34344

[63] Ackerman JN, Zvomuya F, Cicek N, Flaten D. Evaluation of manure-derived struvite as a phosphorus source for canola. Canadian Journal of Plant Science. 2013;**93**(3):419-424

[64] Hertzberger AJ, Cusick RD, Margenot AJ. Maize and soybean response to phosphorus fertilization with blends of struvite and monoammonium phosphate. Plant and Soil. 2021;**461**(1):547-563

[65] Chand S, Paul B, Kumar M. An overview of use of Linz-Donawitz (LD) Steel Slag in Agriculture. Current World Environment. 2015;**10**(3):975-984

[66] Annual Statistics. Strategy paper on resource efficiency in steel sector through recycling of scrap and slag. Niti Aayog. 2017-2018

[67] Das S, Gwon HS, Khan MI, Jeong ST, Kim PJ. Steel slag amendment impacts on soil microbial communities and activities of rice (Oryza sativa L.). Science Report. 2020;**10**:6746

[68] Yildirim IZ, Prezzi M. Chemical, mineralogical and morphological properties of steel slag. Advance Civil Engineering. 2011:143638

[69] Xian W, Qing-Sheng C. Steel slag as an iron fertilizer for corn growth and soil improvement in a pot experiment. Pedosphere. 2006;**16**(4):519-524

[70] White B, Tubana BS, Babu T, Mascagni H Jr, Agostinho F, Datnoff LE, et al. Effect of silicate slag application on wheat grown under two nitrogen rate. Plant. 2017;**6**:47

[71] Ali MA, Oh JH, Kim PJ. Evaluation of silicate iron slag amendment on reducing methane emission from flood water rice farming. Agriculture Ecosystem Environment. 2008;**128**:21-26

[72] Zhang M, Liang Y, Chu G. Applying silicate fertilizer increases both yield and quality of table grape (Vitis vinifera L.) grown on calcareous grey desert soil. Science of Horticulture. 2017;**225**:757-763

[73] Haynes RJ, Belyaeva ON, Kingston G. Evaluation of industrial wastes as sourcesof fertilizer silicon using chemical extractions and plant uptake. Journal of Plant Nutrition and Soil Science. 2013;**176**:238-248

[74] Ning D, Song A, Fafn F, Li Z, Liang Y. Effects of slag-based silicon fertilizer on rice growth and brown-spot resistance. PLoS ONE. 2014;**9**(7):e102681

[75] Ning D, Liang Y, Liu Z, Xiao J, Duan A. Impacts of steel-slag-based silicate fertilizer on soil acidity and silicon availability and metals-immobilization in a paddy soil. PLoS ONE. 2016;**11**(12):e0168163

[76] Gwon HS, Khan MI, Alam MA, Das S, Kim PJ. Environmental risk assessment of steel-making slags and the potential use of LD slag in mitigating methane emissions and the grain arsenic level in rice (Oryza sativa L.). Journal of Hazard Materials. 2018;**353**:236-243

[77] Yang R, Mitchell CC, Howe JA. Relative neutralizing value as an indicator

of actual liming ability of limestone and byproduct materials. Communications in Soil Science and Plant Analysis. 2018;**49**(10):1144-1156

[78] Preston HAF, de Sousa Nuns GH, Preston W, de Souza EB, Mariano RLR, Datnoff LE, et al. Slag-based silicon fertilizer improves the resistance to bacterial fruit blotch and fruit quality of melon grown under field conditions. Crop Protection. 2021;**147**:105460

Chapter 5

Perspective Chapter: Hydroxyapatite – Surface Functionalization to Prevent Bacterial Colonization

Raisha García-Cadme, Irene García Cano, Oscar Castaño and Javier Fernandez

Abstract

Microbial colonization is one of the main causes of implant loosening and rejection. Pathogenic contamination and the subsequent biofilm formation reduce the implant's chance of survival and can be life-threatening to a patient. Among the many strategies employed to reduce the infection probability of bioceramics, surface functionalization plays a key role. This chapter is dedicated to describing the different strategies available to prevent bacterial colonization and the proliferation of hydroxyapatite-coated implants. Moreover, the factors intervening in the bacteria-implant interaction will be described, detailing the mechanisms involved during the contact, adhesion, and proliferation of bacteria. Finally, the characterization methods will be discussed, emphasizing the bioactivity and antibacterial assays.

Keywords: antibacterial, hydroxyapatite, functionalization, bioactivity, implants

1. Introduction

Bioceramics, particularly hydroxyapatite (HA), are used massively to produce ceramic biomaterials and coatings for metallic implants. Implant infection is a serious medical complication and socioeconomic concern.

The economic burden can be quantified by the increased time in the hospital, rehospitalizations, additional surgeries, and the total cost of outpatient care. There are also intangible costs associated with implant infections, such as physical limitation, mental trauma, and reduction in quality of life for the patient.

An infected medical device can be difficult to treat with only antibiotic therapies. When these therapies are ineffective, then required the use of surgical procedures, such as debridement or implant replacement. Once the infection has spread to the bone, known as osteomyelitis, it may lead to limb amputation threatening the patient's life.

The first part of this chapter describes the mechanisms of interaction between the bacteria and the surface of hydroxyapatite. The second part is a review of the different strategies to develop an antibacterial functionalization. The last part describes the main methods of studying bioactivity and antibacterial properties.

2. Bacteria-hydroxyapatite interaction mechanism

Infection of HA implants is a complex problem in clinical medicine. The large number of pathogens that infect medical devices, their resistance to antibiotics, and the strategies that microorganisms used to resist treatments are some of the reasons.

The infection probability is dependent on several aspects, such as the environmental conditions in which the attachment occurs, the type of microorganism, the properties of the substrate, and host characteristics. Furthermore, preexisting conditions, such as diabetes, obesity, and the use of immunosuppressant drugs, can increase the probability of an infection event.

There is limited research about the effect of material composition on the probability of infection. Hailer et al. [1] studied hip implant infection, and they detected no significant difference in infection events comparing hydroxyapatite-coated with micro-rough titanium implants.

The data available show that factors, such as the local immunological environment, the type of surgical intervention, the healing time, the fluid in contact with the implant and the microorganism contained, are more important to determine the infection probability than the chemical composition of the implant [2].

2.1 Implant infection classification

Infections of medical devices can be classified by the timeline of the infection event as early postoperative, late chronic, and hematogenous infection [3]. Alternative classifications are related to the microorganisms detected (bacterial, fungal, and polymicrobial) or by the substrate infected.

Early postoperative infections, also known as surgical site infection (SSI), are associated with postoperative wound infection and nosocomial infections. The symptoms appear during the first three months after surgery because of contamination during implantation or hospitalization before the wound is closed. The most common organism isolated from early infection is *Staphylococcus aureus* [4], but depending on the type of implant and the surrounding tissues, other organisms can also be found, such as *Klebsiella* spp., *Pseudomonas* spp., and *Escherichia coli* [5].

Late chronic infections are delayed postoperative infections and the symptoms normally emerge after the third month and up to two years after the surgery. These infections normally develop after months of apparent implant stability and their treatment includes surgical intervention and implant exchange [6].

Hematogenous infection generally occurs after a symptom-free period and is caused by bacteria originating from a secondary infection that spread through the bloodstream infecting other tissues [7]. This type of infection is a threat to the patient's life years after surgery.

2.2 Bacteria's life cycle

Suitable environmental conditions are necessary for bacterial attachment to the implant surface. During unfavorable or stressful conditions, the bacteria can enter an intermediate or "starvation survival" which allows them to survive long periods of nutrient deprivation [8]. During this stage, the bacteria morphology changes dramatically adapting to a spore-like shape, known as ultramicrobacteria [9], reducing significatively the size and the metabolic activity until the conditions allow for active growth [10].

The bacteria life cycle on a substrate is a process genetically regulated that occurs in four main stages: attachment, growth, proliferation, and dissemination [9, 11]. Starting within the first few seconds after the implant insertion, the bacteria's reversible attachment to the substrate is mediated by Van der Waals forces and determined by the surface charge, the degree of hydration, the topography, and the surface's roughness [12, 13].

An inactive bacterium, also denominated swimmer or planktonic cell, can interact with a surface using a flagellum as a mechanoreceptor. Once a suitable surface is detected, a gene expression allows the bacteria to change phenotypically attaching irreversibly to the substrate [9, 14].

Immediately after the irreversible adhesion, the bacteria begin to grow and proliferate, creating microcolonies of one or several species embedded in an extracellular polymeric matrix or slime [11]. This polymeric slime is composed of exopolysaccharides, proteins, lipids, and extracellular DNA [13, 14]. The biofilm has matrix-enclosed bacterial colonies adherent to each other and surfaces. Biofilms contain open channels that facilitate nutrient and water diffusion from the bulk phase to bacteria in the biofilm [15].

Biofilm generation is a survival strategy that protects bacteria from changes in environmental conditions or antimicrobial agents. Biofilms can protect bacteria by different mechanisms. First, by acting as a barrier that can dilute reactive species before they can reach the bacterial wall. Second, by creating a stationary phase to reduce the effectiveness of antibiotics. Third, by increasing the survival of bacterial subpopulations with antibiotic-resistant phenotypes.

During the initial formation stages, the biofilm is still unstable and susceptible to elimination, but once maturation is achieved it gains an increased thickness (up to 50 μm), a mushroom or column-like morphology, and higher resistance to antibiotics [16]. The dissemination is the culmination of the "bacterial life cycle" by biofilm dissolution and detachment of free-living bacterial cells, which will spread to other locations [9, 16]. **Figure 1** presents a schematic representation of the implant infection timeline.

2.3 Common microorganisms infecting implants

Implants infections are mainly caused by *Staphylococcus* bacteria, the coagulase-positive *S. aureus,* and the coagulase-negative *Staphylococcus epidermidis* [13, 17].

S. aureus can release enzymatic virulence factors, such as exfoliative toxins and nucleases, to avoid the immune response and toxins like hemolysins and leucocidins to destroy host cells [18]. Indeed, considerable concern has recently been raised because of the resistance to antibiotics of methicillin-resistant *Staphylococcus aureus* (MRSA) as the cause of the most important incidents of infectious diseases [19].

In healthy people, *S. epidermidis* is a symbiotic microorganism that inhibits the colonization of more virulent bacteria in the skin and mucous membranes. *S. epidermidis* adheres exceptionally well to indwelling catheters and is founded in early postoperative infections [20].

Other bacterial strains found frequently in early postoperative implant infections are aerobic gram-negative bacilli, such as *Pseudomonas aeruginosa* and *Escherichia coli*. These microorganisms are especially threatening due to their high virulence and resistance to antimicrobial agents, mainly because they can produce a mature biofilm in 5–7 days [21]. *P. aeruginosa*, an opportunistic pathogen, found in soil or water is one of the major causes of nosocomial infections [22]. The pathogenesis of *P. aeruginosa*

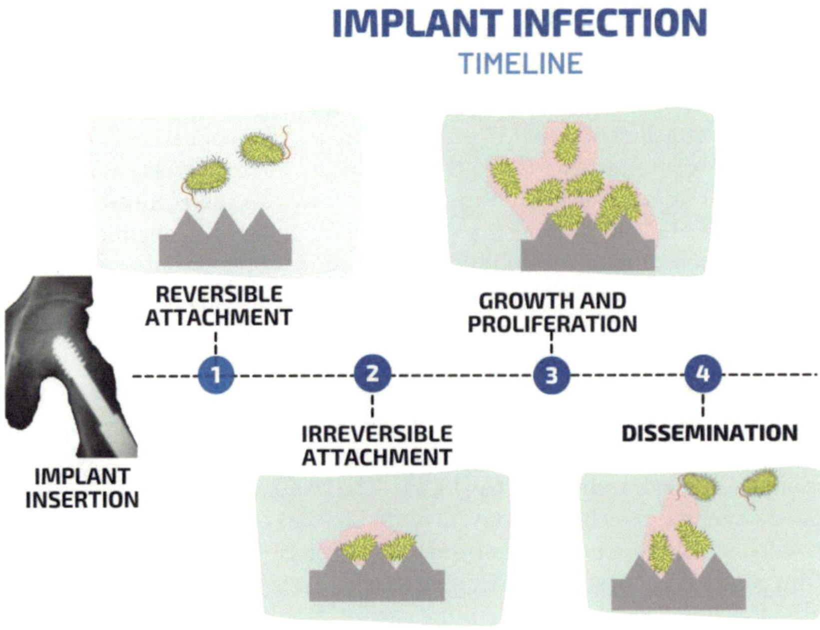

Figure 1.
Implant infection timeline. The Figure created by author using parts of figures from Servier Medical Art, provided by Servier, licensed under a Creative Commons Attribution 3.0 unported license.

is related to their high adherence to mucus and injured-epithelial cells [23]. *E. coli* is a symbiotic organism found in the gut flora of healthy people and can cause more than 80% of urinary tract infections (UTI). *E. coli* is the second highest cause of gram-negative orthopedic implant infection [22].

Other microorganisms found infecting orthopedic devices are *Streptococcus* and *Enterobacteriaceae* strains [7, 24]. If the implant is in direct contact with mucus or sores on the skin *Streptococcus pyogenes*, *Streptococcus agalactiae,* and *Streptococcus pneumoniae* can also be present [18, 25].

Polymicrobial infections occur generally in the early stages after surgery but can be discovered months afterward in immunocompromised patients. When two or more microbes colonize the same substrate, they interact releasing small molecules that change the host's environment. These biomolecules can increase the proliferation of the microbes improving their resistance to antimicrobial drugs resulting in a challenge for treatment [26, 27].

Fungal infections of implants are rare events, but their treatment involves more medical problems and aggressive surgical treatments than normal bacterial infections. Most fungal infections are caused by candida species, such as *Candida albicans* and *Candida parapsilosis* [28, 29].

3. Mechanisms against bacterial infections

Meticulous aseptic methods during surgery and prophylactic actions, such as antibiotic treatment, are not enough to prevent an implant infection. Several strategies can be used to improve the resistance to an infection event. These strategies can be broadly differentiated by the production of an implant with intrinsic antibacterial properties and by the functionalization of the surface.

3.1 Intrinsic antibacterial properties of HA

Intrinsic antibacterial properties are dependent on the chemical composition, structure, and morphology of the material. The obtention of HA with intrinsic antibacterial properties can be achieved by multi-cationic/anionic substitution, also known as ion exchange [30].

Substitutions of calcium ions, phosphate, and hydroxyl groups in the HA lattice are possible because the HA-crystal can incorporate several elements with different atomic radii and charges [31]. Cations, such as copper, iron, magnesium, manganese, potassium, strontium, and zinc, among others, can substitute some calcium cations. Similarly, anions, such as chlorine, fluorine, and carbonate groups, can substitute hydroxyl and phosphate lattice positions.

The exchange of atoms opens a wide range of possibilities for customizing the properties of HA. The atomic substitution alters the solubility, reactivity, and biological properties of HA. Also, the amount of substituting atoms modify the percentage of the amorphous/crystalline phase ratio affecting the dissolution rate and the duration of the antibacterial properties. A co-doping strategy has been used to stabilize the crystalline structure after the introduction of atoms with a very big difference in radii and charge [32–35]. The objective is to improve bioactivity while maintaining good antibacterial properties.

There is a wide range of transition metals exhibiting antimicrobial activity being eligible for the cationic substitution of HA. Some examples are silver [33, 36–39], zinc [40–42], copper [34, 43, 44], and gallium [45–47]. The antibacterial characteristic of transition metals is produced by their oxidated forms [48]. The mechanism proposed is based on the inhibition of enzymes and cytoplasmatic proteins by the reaction with electron donors, creating M-thiolate bonds [49]. The inactivation of cytoplasmatic proteins can produce the disturbance of membrane potential, increasing permeability and the leaking of cellular contents [50].

Metallic ions can induce oxidative stress by increasing the production of reactive oxygen species (ROS) via the Fenton and Haber–Weiss reaction. These reactive oxygen species can react with DNA molecules and proteins, triggering condensation and denaturalization reactions, reducing the replication capacity [51]. **Figure 2** illustrates the main mechanisms of ROS toxicity.

The synthesis of multi-substituted HA can be achieved by different methods, such as sol-gel synthesis [35], co-precipitation [37, 41, 52, 53], hydrothermal [54, 55], and ball-milling [56, 57], among others [39, 42, 58–61].

The main advantage of the use of atomic substitution to increase the antibacterial resistance of implants fabricated with hydroxyapatite is the straightforward adaptation of the production facilities to prepare substituted HA. The atomic substitution can be achieved either during the synthesis of the HA or during the posttreatment of pure HA.

The challenge in the use of intrinsic antibacterial properties is related to the precise control of composition and percentage of the crystalline/amorphous phases. These variables simultaneously regulate the solubility and the release control of the antibacterial compound.

It is necessary to select the best variable combination to ensure long-term antibacterial properties and optimal concentration to avoid toxicity problems. The

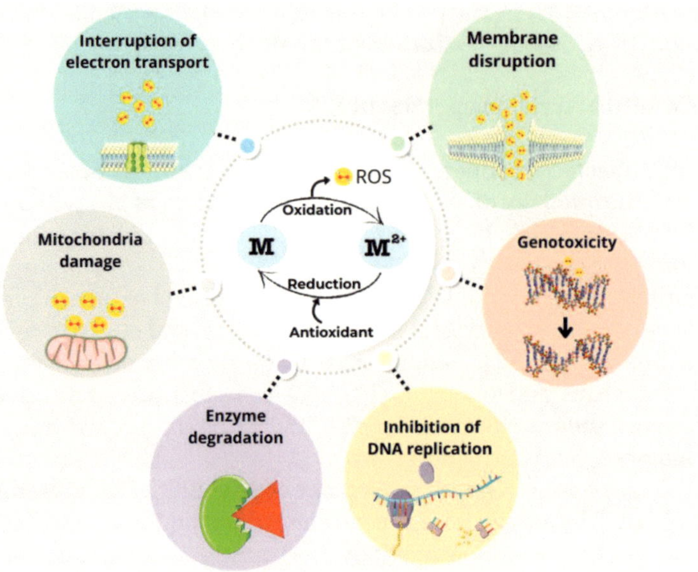

Figure 2.
Mechanisms of ROS toxicity. The Figure created by author using parts of figures from Servier Medical Art, provided by Servier, licensed under a Creative Commons Attribution 3.0 unported license.

compositional adjustment is an interesting strategy, combining different solubility rates and releasing kinetics to obtain antibacterial properties consistently during a long period of time [59]. **Figure 3** presents a schematic representation of Projection of Ag-substituted HA according to planes 110, 001, and 111.

3.2 Antibacterial functionalization of HA

Functionalization can be defined as a modification of the material surface by incorporating functional groups, biomolecules, nanoparticles, and other components with the objective to modify or enhance properties. The surface functionalization is mainly used to increase bioactivity, osseointegration, and angiogenesis but also can be used to produce antibacterial coatings on HA. Functionalization can be achieved by linking biomolecules by a covalent bond to the HA surface, by using physical adsorption to generate a covering layer, or by creating a hybrid coating.

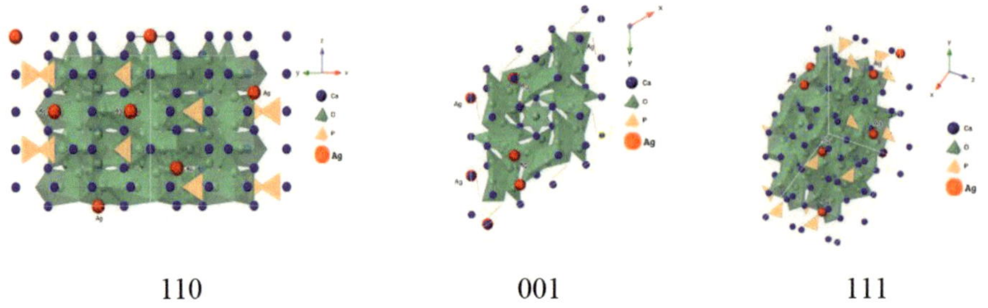

Figure 3.
Projection of Ag-substituted HA according to planes 110, 001, and 111. Figure created by author.

3.2.1 Functionalization by covalent bonding

A covalent bond can be created because the HA surface presents approximately 2.6 P-OH groups per nm^2 that can be used as an anchoring point to tether a molecule to the surface of HA via hydrogen bonding [62]. The target molecule can be covalently anchored if containing a functional group as amines, carboxylates, and thiols that can react with the hydroxyl moieties [63–65].

Another approach is to use a molecular adhesive as a spacer between the HA surface and the antibacterial compound. Molecular adhesives act like a bridge, creating a covalent bond between the hydroxyapatite surface and the biomolecule of interest [66, 67]. Many molecular adhesives can produce a covalent link with HA, but the most used are silane coupling agents that contain a reactive silanol group at one end and at the other end a hydrolysable group, typically alkoxy, halogen, or amine [68]. CEPTES and APTES are examples of silane molecules that contain a reactive group of carboxyl and amino, respectively, that can react with a functional group of a target biomolecule [65, 69, 70].

The main drawback of the use of silane agents as adhesive molecules is their high reactivity. Silanes interact through hydrogen bonding with the hydroxyl groups of the surface, but lateral polymerization may occur, generating multiple siloxane layers. The creation of a monolayer of silane molecules at the HA surface normally requires anhydrous conditions, extended reaction times, elevated temperatures (50–120°C), and rigorous control of the reagent concentration [68].

Formulated bio-adhesives using mussel adhesive proteins (MAPs) are a good alternative for silane coupling agents [71, 72]. Many MAPs have been isolated containing L-3,4-dihydroxyphenylalanine (Dopa) residues. The functional part of Dopa residues necessary to create covalent bonds is the catechol group.

A catechol group is composed of a benzene ring and two hydroxyls in the ortho position. Catechol groups can be oxidated into quinones under alkaline and neutral conditions, creating a stable coating with controllable film thickness [73, 74]. Dopamine, caffeic acid, and L-3,4-dihydroxyphenylalanine have been used to produce stable coatings containing the molecule of interest [75–77].

3.2.2 Functionalization by physical adsorption

Physical adsorption is a straightforward method to generate antibacterial functionalization because it can be achieved by soaking HA powders or substrates into a solution containing the antibacterial molecule. The main drawback is the weak bonds or interactions resulting in high release kinetics of the adsorbed molecule.

The nature and chemistry of the HA control the interaction with biomolecules, and the surface chemistry and porosity are fundamental variables. The interaction between adsorbed molecules and hydroxyapatite surfaces is predominantly Van der Waals forces, hydrogen, and weak coordination bonds [78].

The adsorption capacity is dependent on the surface area available. When there is no internal porosity, the biomolecules are mostly adsorbed onto the surface, limiting the loading capacity. Mesoporous hydroxyapatites are excellent structures for biomolecule release using physical adsorption mainly by their high surface area, but also because the pore size and interconnection can be customized [65, 79].

The chemical composition of the HA surface can be functionalized to improve adsorption and reduce the releasing kinetics prolonging the duration and efficiency of the antibacterial properties (**Figure 4**) [80, 81].

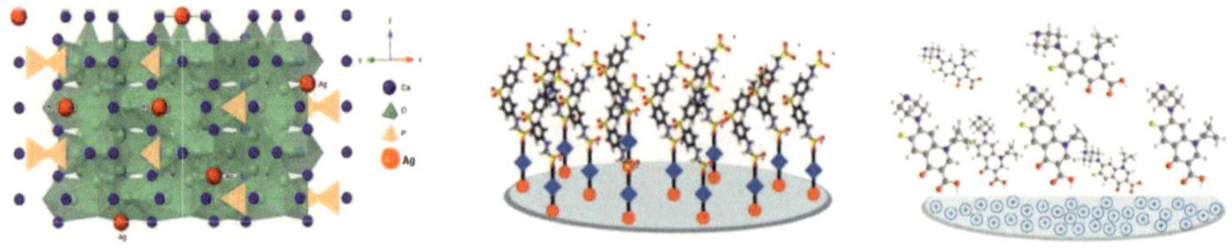

Figure 4.
Strategies to produce antibacterial properties on HA. Figure created by author.

3.2.3 Functionalization by antibacterial composites

The production of antibacterial composites is a valuable approach to produce antibacterial functionalization of hydroxyapatite. The combination of polymers, nanoparticles, antibiotics, and different structures of HA have unlimited possibilities for the obtention of medical devices with antibacterial properties.

Among the methods used to prepare composite, the incorporation of antibacterial compounds into the precursor solution is highly employed in co-precipitation, sol-gel, and hydrothermal methods to produce antibacterial HA powders with high homogeneity [42, 82, 83]. Additionally, the solution containing the antibacterial compound can be applied to HA substrates by dip coating and spin coating [84–86]. Another approach is to use electrochemical methods, such as electrodeposition [34, 87–89], micro-arc oxidation [90], and electrophoretic deposition [91], to generate HA coatings containing antimicrobial compounds. Electrospinning can be used for the creation of composite nanofibers and coatings with a combination of functionalized-HA nanoparticles and antimicrobial compounds [92–94]. Other methodologies include the spraying of a solution containing the precursors of HA mixed with the antibacterial compound [95–97].

Different compounds can be used to produce antibacterial functionalization of HA including antibacterial polymers, nanoparticles, antibiotics, and proteins. Among them, antibacterial polymers, such as chitosan, and polycaprolactone (PCL), are frequently preferred for the obtention of HA composites with antibacterial properties [87, 90, 91, 98–102]. Metallic nanoparticles, such as silver, zinc dioxide, titanium dioxide, and niobium pentoxide, have been combined with graphene and hydroxyapatite to produce composites with antibacterial properties [64, 80, 94, 103–105].

Many antibiotics have been used to produce antibacterial functionalization. Among them, β-lactam antibiotics, such as amoxicillin, can be adsorbed onto HA nanocrystals to be used alone or in composites fabricated by electrospinning [92, 106].

Fluoroquinolones, such as ciprofloxacin, have been used to functionalize nano-HA crystals and composites of chitosan, poly(vinyl alcohol), and HA sponges [107, 108]. Other fluoroquinolones, such as enoxacin, have been employed together with nano-HA/Polyurethane-cement to enhance the antibacterial properties of bone cement [109]. Adsorption has been used to introduce broad-spectrum antibiotics of the tetracycline class and cyclic oligosaccharides, such as cyclodextrin, onto HA surfaces [78, 88, 95]. Moreover, antibiotics from the glycopeptides class as vancomycin has been loaded into porous substrates to obtain a controlled release [84, 91]. Taha et al. [95] prepared a cyclodextrin polymer loaded with rifampicin, an antimycobacterial antibiotic, to prepare a coating onto titanium-coated hydroxyapatite surfaces. Covalent immobilization of antibiotics like doxorubicin can be achieved using HA nanoparticles functionalized with amino groups or by more complicated methods, including the fabrication of polymer brushes, to anchor aminoglycoside antibiotics, such as gentamycin, to HA substrates [110, 111]. Furthermore, aminoglycosides antibiotics, such as streptomycin, have been encapsulated in HA nanoparticles [93].

Proteins that present a broad spectrum of antibacterial activity can be used for the antibacterial functionalization of HA surfaces. One example is protamine, a cationic protein rich in arginine residues, used for Koizumi et al. [112] to functionalize different calcium phosphates by adsorption. Peptides with between 10 and 15 amino acids are preferable for the antibacterial functionalization of surfaces because of the lower production cost compared to proteins [113]. These peptides so-called antibacterial peptides (AMPs) are very interesting due to their selectivity and high antibacterial efficiency at low concentrations. Their efficacy is based on the adoption of amphipathic structures and their cationic character. So far, more than 700 types of AMP have been isolated from different organisms. AMPs can be used to functionalize HA surfaces by electrostatic and covalent attachment [114].

4. Characterization methods

4.1 Bioactivity

One of the most important factors affecting the bioactivity and biocompatibility of HA implants is the release of substances that can cause toxicity, hypersensitivity, allergies, or even osteolysis depending on the released product, their concentration, and the exposure time [115]. Ensuring a controlled release of substances after the implantation is one of the key strategies to improve the implant performance, as it can affect osseointegration and implant long-term viability. The ISO 10993-17 is the standard that establishes the limits for leachable substances in medical devices.

Immersion tests are used to quantify the products released at body temperature (37 ± 1°C) under static or dynamic conditions. The level of substance released is mainly dependent on the implant's surface area and the composition of both implant and the body fluid in contact. Therefore, the released products should be determined using a solution with the closest composition to the body fluid in contact with the implant under working conditions. Complex biofluids can be replicated with phosphate buffer saline (PBS), Hank's solution, simulated body fluid (SBF), Ringer's solution, artificial saliva, and eagle's minimum essential medium (EMEM), as well as other fluids [116, 117].

The ability of an orthopedical implant to induce the formation of biological apatite on its surface is one of the requirements to determine correct osseointegration. The precipitation of apatite can be replicated *in vitro* by the immersion of the sample in simulated body fluid, a saturated solution with a composition comparable to the human blood plasma [118].

The cytotoxicity of an antibacterial functionalization is determined by cell viability and cell proliferation. Cell viability refers to the number of live, healthy cells in a sample and cell proliferation is defined by the valance between cell divisions and cell loss through cell death or differentiation. Cell viability assays can be used to evaluate cell health and can be assessed by culturing the chosen cells over either the sample or an extraction vehicle. The cell viability can be quantified using redox indicators that interact with metabolites produced by healthy cells. Another method is the use of dyes that react only with healthy cells like methylene blue, triptan blue, neural red, or by live/dead assays [119–121].

Cell proliferation is defined as the increase of cell number after the final step of the cell cycle due to cytokinesis or cell division. Many strategies can be used to assess cellular proliferation as the use of nucleoside-analogs incorporated during DNA synthesis, the quantification of cell cycle-associated proteins, and the use of cytoplasmatic proliferation dyes [122–125].

All methods are valid to compare proliferation but it is important to consider their strength and limitations and, to improve the accuracy of the results, multiple assays should be performed [126]. The standard procedure of in vitro cell viability and proliferation assays is exposed in ISO 10993-5: Biological evaluation of medical devices-tests for in vitro cytotoxicity.

4.2 Antibacterial properties

The antibacterial properties of a biomaterial can be tested by studying antimicrobial susceptibility *in vitro*. Multiple methods can be used to evaluate, either quantitatively or quantitatively, the antibacterial activity of HA coatings and powders.

Qualitative measurements may not provide quantifiable results but offer valuable information regarding the bacteria's sensitivity to antimicrobial functionalization of materials.

- The Kirby–Bauer disk diffusion susceptibility test, also known as the agar disk diffusion method, is a standardized procedure to qualitatively determine the sensitivity or resistance of bacteria to antimicrobial compounds [127]. The presence or absence of growth around the disk is an indirect measure of the bacterial inhibition by the antimicrobial compound. The Kirby–Bauer test was designed to test antibiotic-impregnated disks, but many authors have also used it to test antibacterial substrates. This method cannot be used to determine the minimum inhibitory concentration (MIC) but can be approximate for some microorganisms and antibiotics by comparing the inhibition zone using systems that can read and interpret the results [128]. The main advantages of this method are its simplicity and low cost.

Quantitative tests provide more accurate information about bacterial growth in presence of an antimicrobial compound. These methods are normally based on the measurement of the turbidity of a bacterial solution to indirectly assess the bacteria's sensitivity to an antibacterial compound.

- Among the quantitative test used, the broth dilution test can be used to test both coatings and powders that release the antibacterial compound. This method is based on the preparation of dilutions of the antibiotic or the extract in a liquid growth medium. The dilutions are inoculated with a previously known concentration of bacterial suspension. After overnight incubation, the turbidity is measured, and the MIC is defined as the lowest concentration that prevented the growth of the microorganism.

- Once the MIC is determined, it may be useful to determine the interaction of the antibacterial compound depending on the time. The time-kill assay is based on the preparation of antimicrobial extracts with dilutions lower than MIC, and up to 16 x MIC that is inoculated with the same concentration of bacterial suspension, and their growth is measured during different intervals of time [129].

Likewise, in any biological test, the results obtained from a bacterial sensitivity test are dependent on variables, such as the inoculum size, the type of growth medium, and the incubation time [130]. Updated standards should be used to obtain reliable results.

Among the standardized methods to quantitatively evaluate antibacterial activity, there is the ASTM E2149-standard test method for determining the antimicrobial activity of antimicrobial agents under dynamic contact conditions and ASTM E2180-standard test method for determining the activity of incorporated antimicrobial agent in polymeric or hydrophobic materials.

5. Conclusions

This book chapter presented the mechanisms that bacteria use to attach and proliferate on implants. Moreover, the main strategies used to provide antibacterial properties to hydroxyapatite powders and substrates were exposed.

The obtention of medical devices with suitable antibacterial properties must be complemented by excellent biocompatibility and adequate mechanical properties. Novel strategies include the combination of different methodologies and the use of different compounds to improve the properties.

The main difficulty in developing antibacterial functionalization of implants is the lack of homogeneity in the in vitro assays, which limits the comparison of the strategies employed. Additionally, many factors can affect the results from in vitro assays, such as the type of cell, their origin, incubation time, and the compound used to quantify the proliferation.

To accelerate the development of suitable antibacterial functionalization, more efforts must be made to use standardized protocols for bioactivity and antibacterial in vitro assays. The homogenization of the assays is necessary for an accurate comparison of the release of substances and bioactivity. Furthermore, through an in-depth study of the antibacterial properties during long periods of time, a selection of the suitable strategy for each application can be made.

Even though many attempts have been made to produce antibacterial functionalization of HA, none have been used industrially. Considerably work still needs to manufacture a cost-effective implant's antibacterial functionalization.

Author details

Raisha García-Cadme*, Irene García Cano, Oscar Castaño and Javier Fernandez
University of Barcelona, Barcelona, Spain

*Address all correspondence to: raisha.garcia@ub.edu

References

[1] Hailer NP, Lazarinis S, et al. Hydroxyapatite coating does not improve uncemented stem survival after total hip arthroplasty! Acta Orthopaedica. 2015;**86**(1):18-25. DOI: 10.3109/17453674.2014.957088

[2] Yue C, Zhao B, et al. The implant infection paradox: Why do some succeed when others fail? European Cells Materials. 2015;**29**:303-313. DOI: 10.22203/eCM.v029a23

[3] Tsukayama DT, Estrada R, et al. Infection after total hip arthroplasty. Journal of Bone Joint Surgery. 1996;**78**(4):512-523. DOI: 10.2106/00004623-199604000-00005

[4] Jain S, Chowdhury R, et al. Characterization of the clonal profile of methicillin resistant Staphylococcus aureus isolated from patients with early post-operative orthopedic implant based infections. Annals of Clinical Microbiology and Antimicrobials. 2019;**18**(8):1-7. DOI: 10.1186/s12941-019-0307-z

[5] Kumar S, Sengupta M, et al. Early post-operative wound infection in patients undergoing orthopaedic surgery with implant. Internationa Journal of Science Study. 2017;**5**(8):44-48. DOI: 10.17354/ijss/2017/518

[6] Sloan BH, McNab AA, et al. Late infection of hydroxyapatite orbital implants. Australian and New Zealand Journal of Ophthalmology. 1999;**27**(1): 45-48. DOI: 10.1046/j.1440-1606.1999.00160.x

[7] Zeller V, Kerroumi Y, et al. Analysis of postoperative and hematogenous prosthetic joint-infection microbiological patterns in a large cohort. The Journal of Infection. 2018;**76**(4):328-334. DOI: 10.1016/j.jinf.2017.12.016

[8] Klančnik A, Guzej B, et al. Stress response and pathogenic potential of campylobacter jejuni cells exposed to starvation. Research in Microbiology. 2009;**160**(5):345-352. DOI: 10.1016/j.resmic.2009.05.002

[9] Hall-Stoodley L, Costerton JW, et al. Bacterial biofilms: From the natural environment to infectious diseases. Nature Reviews. Microbiology. 2004;**2**:95-108. DOI: 10.1038/nrmicro821

[10] Hu Y, Coates A. Nonmultiplying bacteria are profoundly tolerant to antibiotics. In: ARM C, editor. Antibiotic Resistance. 1st ed. Berlin, Heidelberg: Springer Berlin Heidelberg; 2012. pp. 99-119. DOI: 10.1007/978-3-642-28951-4_7

[11] Deva AK, Adams WP, et al. The role of bacterial biofilms in device-associated infection. Plastic and Reconstructive Surgery. 2013;**132**(5):1319-1328. DOI: 10.1097/PRS.0b013e3182a3c105

[12] Singh AV, Vyas V, et al. Quantitative characterization of the influence of the nanoscale morphology of nanostructured surfaces on bacterial adhesion and biofilm formation. Bansal V, editor. PLoS One. 2011;**6**(9):e25029. DOI: 10.1371/journal.pone.0025029

[13] Arciola CR, Campoccia D, et al. Biofilm formation in staphylococcus implant infections. A review of molecular mechanisms and implications for biofilm-resistant materials. Biomaterials. 2012;**33**(26):5967-5982. DOI: 10.1016/j.biomaterials.2012.05.031

[14] O'Neill E, Pozzi C, et al. A novel Staphylococcus aureus biofilm phenotype mediated by the fibronectin-binding proteins, FnBPA and FnBPB. Journal of Bacteriology. 2008;**190**(11):3835-3850. DOI: 10.1128/JB.00167-08

[15] Wilking JN, Zaburdaev V, et al. Liquid transport facilitated by channels in Bacillus subtilis biofilms. Proceedings of the National Academy of Sciences. 2013;**110**(3):848-852. DOI: 10.1073/pnas.1216376110

[16] Webb JS, Givskov M, et al. Bacterial biofilms: Prokaryotic adventures in multicellularity. Current Opinion in Microbiology. 2003;**6**(6):578-585. DOI: 10.1016/j.mib.2003.10.014

[17] Oliveira WF, Silva PMS, et al. Staphylococcus aureus and Staphylococcus epidermidis infections on implants. The Journal of Hospital Infection. 2018;**98**(2):111-117. DOI: 10.1016/j.jhin.2017.11.008

[18] Stoneham S, Peters J, et al. Staphylococcal and streptococcal infections. Medicine (Baltimore). 2021;**49**(12):731-738. DOI: 10.1016/j.mpmed.2021.09.001

[19] Lee AS, de Lencastre H, et al. Methicillin-resistant Staphylococcus aureus. Nature Review Disorders Primers. 2018;**4**(1):18033. DOI: 10.1038/nrdp.2018.33

[20] Fey PD, Olson ME. Current concepts in biofilm formation of Staphylococcus epidermidis. Future Microbiology. 2010;**5**(6):917-933. DOI: 10.2217/fmb.10.56

[21] Høiby N, Ciofu O, et al. The clinical impact of bacterial biofilms. International Journal of Oral Science. 2011;**3**(2):55-65. DOI: 10.4248/IJOS11026

[22] Hsieh P, Lee MS, et al. Gram-negative prosthetic joint infections: Risk factors and outcome of treatment. Clinical Infectious Diseases. 2009;**49**(7):1036-1043. DOI: 10.1086/605593

[23] Delden C. Virulence factors in Pseudomonas aeruginosa. In: Ramos JL, editor. Virulence and Gene Regulation. 1st ed. Boston, MA: Springer, Boston, MA; 2004. pp. 3-45. DOI: 10.1007/978-1-4419-9084-6_1

[24] Campoccia D, Montanaro L, et al. The significance of infection related to orthopedic devices and issues of antibiotic resistance. Biomaterials. 2006;**27**(11):2331-2339. DOI: 10.1016/j.biomaterials.2005.11.044

[25] Loubet P, Koumar Y, et al. Clinical features and outcome of Streptococcus agalactiae bone and joint infections over a 6-year period in a French university hospital. Ghozy S, editor. PLoS One. 2021;**16**(3):e0248231. DOI: 10.1371/journal.pone.0248231

[26] Marculescu CE, Cantey RJ. Polymicrobial prosthetic joint infections: Risk factors and outcome. Clinical Orthopaedics and Related Research. 2008;**466**(6):1397-1404. DOI: 10.1007/s11999-008-0230-7

[27] Vila T, Kong EF, et al. Therapeutic implications of C. albicans-S. aureus mixed biofilm in a murine subcutaneous catheter model of polymicrobial infection. Virulence. 2021;**12**(1):835-851. DOI: 10.1080/21505594.2021.1894834

[28] Ponde NO, Lortal L, et al. Candida albicans biofilms and polymicrobial interactions. Critical Reviews in Microbiology. 2021;**47**(1):91-111. DOI: 10.1080/1040841X.2020.1843400

[29] Karczewski D, Ren Y, et al. Candida periprosthetic joint infections — Risk factors and outcome between albicans and non-albicans strains. International Orthopaedics. 2022;**46**(3):449-456. DOI: 10.1007/s00264-021-05214-y

[30] Kolmas J, Groszyk E, et al. Substituted hydroxyapatites with antibacterial properties. BioMed

Research International. 2014;**2014**:1-15. DOI: 10.1155/2014/178123

[31] Tabassum S. Role of substitution in bioceramics. In: Khan S, Chaudhry AA, editors. Handbook of Ionic Substituted Hydroxyapatites. Cambridge: Elsevier; 2020. pp. 117-148. DOI: 10.1016/C2018-0-00399-0

[32] Huang Y, Qiao H, et al. Improving the bioactivity and corrosion resistance properties of electrodeposited hydroxyapatite coating by dual doping of bivalent strontium and manganese ion. Surfaces Coatings Technology. 2016;**291**:205-215. DOI: 10.1016/j.surfcoat.2016.02.042

[33] zhang X, Wang B, et al. Chemical stability, antibacterial and osteogenic activities study of strontium-silver co-substituted fluorohydroxyapatite nanopillars: A potential multifunctional biological coating. Ceramics International 2020;**46**(17):27758-27773. DOI:10.1016/j.ceramint.2020.07.275

[34] Karthika A. Biocompatible iron and copper incorporated nanohydroxyapatite coating for biomedical implant applications. Mater Today Proceedings. 2022;**51**(4):1754-1759. DOI: 10.1016/j.matpr.2020.12.813

[35] Samani S, Hossainalipour SM, et al. In vitro antibacterial evaluation of sol-gel-derived Zn-, Ag-, and (Zn + Ag)-doped hydroxyapatite coatings against methicillin-resistant Staphylococcus aureus. Journal of Biomedial Materials Research Part A. 2013;**101A**:222-230. DOI: 10.1002/jbm.a.34322

[36] Ciobanu CS, Iconaru SL, et al. Synthesis and antimicrobial activity of silver-doped hydroxyapatite nanoparticles. BioMed Research International. 2013;**2013**:1-10. DOI: 10.1155/2013/916218

[37] Iconaru SL, Predoi D, et al. Development of silver doped hydroxyapatite thin films for biomedical applications. Coatings. 2022;**12**(3):341. DOI: 10.3390/coatings12030341

[38] Wilcock CJ, Stafford GP, et al. Preparation and antibacterial properties of silver-doped nanoscale hydroxyapatite pastes for bone repair and augmentation. Journal of Biomedical Nanotechnology. 2017;**13**(9):1168-1176. DOI: 10.1166/jbn.2017.2387

[39] Koizhaiganova M, Yasa I, et al. Characterization and antimicrobial activity of silver doped hydroxyapatite obtained by the microwave method. Materials Science. 2016;**22**(3):403-408. DOI: 10.5755/j01.ms.22.3.12133

[40] Peñaflor Galindo TG, Kataoka T, et al. Preparation of nanocrystalline zinc-substituted hydroxyapatite films and their biological properties. Colloid Interface Science Communications. 2016;**10-11**: 15-19. DOI: 10.1016/j.colcom.2016.03.005

[41] Thian ES, Konishi T, et al. Zinc-substituted hydroxyapatite: A biomaterial with enhanced bioactivity and antibacterial properties. Journal of Materials Science. Materials in Medicine. 2013;**24**:437-445. DOI: 10.1007/s10856-012-4817-x

[42] Iqbal N, Iqbal S, et al. Zinc-doped hydroxyapatite—Zeolite/polycaprolactone composites coating on magnesium substrate for enhancing in-vitro corrosion and antibacterial performance. Transactions of the Nonferrous Metals Society of China. 2020;**30**(1):123-133. DOI: 10.1016/S1003-6326(19)65185-X

[43] Unabia RB, Candidato RT, et al. In vitro studies of solution precursor plasma-sprayed copper-doped hydroxyapatite coatings with increasing copper content. Journal of Biomedical Materials Research Part B: Applied

Biomaterials. 2020;**108**(6):2579-2589. DOI: 10.1002/jbm.b.34589

[44] Lyasnikova AV, Markelova OA, et al. Comprehensive characterization of plasma-sprayed coatings based silver- and copper-substituted hydroxyapatite. Powder Metallurgy and Metal Ceramics. 2016;**55**(5-6):328-333. DOI: 10.1007/s11106-016-9809-9

[45] Mellier C, Fayon F, et al. Design and properties of novel gallium-doped injectable apatitic cements. Acta Biomaterialia. 2015;**24**:322-332. DOI: 10.1016/j.actbio.2015.05.027

[46] Melnikov P, Teixeira AR, et al. Gallium-containing hydroxyapatite for potential use in orthopedics. Materials Chemistry and Physics. 2009;**117**(1):86-90. DOI: 10.1016/j.matchemphys.2009.05.046

[47] Kurtjak M, Vukomanović M, et al. Antibacterial nanocomposite of functionalized nanogold and gallium-doped hydroxyapatite. Materials Letters. 2017;**193**:126-129. DOI: 10.1016/j.matlet.2017.01.092

[48] Xiu Z ming, Zhang Q bo, et al. Negligible particle-specific antibacterial activity of silver nanoparticles. Nano Letters 2012;**12**(8):4271-4275. DOI:10.1021/nl301934w

[49] Gordon O, Vig Slenters T, et al. Silver coordination polymers for prevention of implant infection: Thiol interaction, impact on respiratory chain enzymes, and hydroxyl radical induction. Antimicrobial Agents and Chemotherapy. 2010;**54**(10):4208-4218. DOI: 10.1128/AAC.01830-09

[50] Feng QL, Wu J, et al. A mechanistic study of the antibacterial effect of silver ions on Escherichia coli and Staphylococcus aureus. Journal of Biomedical Materials Research. 2000;**52**(4):662-668. DOI: 10.1002/1097-4636(20001215)52:4<662::AID-JBM10>3.0.CO;2-3

[51] Valko M, Morris H, et al. Metals, toxicity and oxidative stress. Current Medicinal Chemistry. 2005;**12**(10):1161-1208. DOI: 10.2174/0929867053764635

[52] Lyasnikova AV, Dudareva OA, et al. Comparative analysis of plasma bioceramic coatings based on zinc-substituted hydroxyapatite and tricalcium phosphate. Glass and Ceramics. 2018;**75**(3-4):163-167. DOI: 10.1007/s10717-018-0048-z

[53] Barbosa AA, Júnior SA, et al. Multifunctional hydroxyapatite with potential for application in theranostic nanomedicine. Materials Science and Engineering: C. 2020;**116**:111227. DOI: 10.1016/j.msec.2020.111227

[54] Li M, Xiao X, et al. Structural characterization of zinc-substituted hydroxyapatite prepared by hydrothermal method. Journal of Materials Science. Materials in Medicine. 2008;**19**:797-803. DOI: 10.1007/s10856-007-3213-4

[55] Ortiz IY, Raybolt dos Santos A, et al. In vitro assessment of zinc apatite coatings on titanium surfaces. Ceramics International. 2016;**42**(14):15502-15510. DOI: 10.1016/j.ceramint.2016.06.203

[56] Vu AA, Robertson SF, et al. Mechanical and biological properties of ZnO, SiO2, and Ag2O doped plasma sprayed hydroxyapatite coating for orthopaedic and dental applications. Acta Biomaterialia. 2019;**92**:325-335. DOI: 10.1016/j.actbio.2019.05.020

[57] Ke D, Robertson SF, et al. Effects of MgO and SiO2 on plasma-sprayed hydroxyapatite coating: An in vivo study in rat distal femoral defects. ACS Applied

Materials & Interfaces. 2017;**9**(31):25731-25737. DOI: 10.1021/acsami.7b05574

[58] Wang JY, Liu YC, et al. Flame-sprayed strontium- and magnesium-doped hydroxyapatite on titanium implants for osseointegration enhancement. Surfaces Coatings Technology. 2020;**386**:125452. DOI: 10.1016/j.surfcoat.2020.125452

[59] Ke D, Vu AA, et al. Compositionally graded doped hydroxyapatite coating on titanium using laser and plasma spray deposition for bone implants. Acta Biomaterialia. 2019;**84**:414-423. DOI: 10.1016/j.actbio.2018.11.041

[60] Hidalgo-Robatto BM, López-Álvarez M, et al. Pulsed laser deposition of copper and zinc doped hydroxyapatite coatings for biomedical applications. Surfaces Coatings Technology. 2018;**333**:168-177. DOI: 10.1016/j.surfcoat.2017.11.006

[61] Bir F, Khireddine H, et al. Electrochemical depositions of fluorohydroxyapatite doped by Cu2+, Zn2+, Ag+ on stainless steel substrates. Applied Surface Science. 2012;**258**(18):7021-7030. DOI: 10.1016/j.apsusc.2012.03.158

[62] Tanaka H, Futaoka M, et al. Structure of synthetic calcium hydroxyapatite particles modified with pyrophosphoric acid. Journal of Colloid and Interface Science. 2005;**283**(2):609-612

[63] Yewle JN, Wei Y, et al. Oriented immobilization of proteins on hydroxyapatite surface using bifunctional bisphosphonates as linkers. Biomacromolecules. 2012;**13**(6):1742-1749

[64] Lazić V, Smičiklas I, et al. Antibacterial ability of supported silver nanoparticles by functionalized hydroxyapatite with 5-aminosalicylic acid. Vacuum. 2018;**148**:62-68

[65] Sistanipour E, Meshkini A, et al. Catechin-conjugated mesoporous hydroxyapatite nanoparticle: A novel nano-antioxidant with enhanced osteogenic property. Colloids Surfaces B Biointerfaces. 2018;**169**:329-339. DOI: 10.1016/j.colsurfb.2018.05.046

[66] Hu H, Wang H, et al. Synthesis and characterization of a PAMAM dendrimer nanocarrier functionalized by HA for targeted gene delivery systems and evaluation in vitro. Journal of Biomaterials Science. Polymer Edition. 2021;**32**(2):205-228. DOI: 10.1080/09205063.2020.1827921

[67] Damia C, Marchat D, et al. Functionalization of phosphocalcic bioceramics for bone repair applications. Materials Science and Engineering: C. 2019;**95**:343-354. DOI: 10.1016/j.msec.2018.01.008

[68] Mittal KL. Silanes and Other Coupling Agents, Volume 4. 1st Ed. Vol. 5, CRC Press Book. London: Brill; 2007. p. 348. DOI: 10.1201/b12244

[69] Russo L, Taraballi F, et al. Carbonate hydroxyapatite functionalization: A comparative study towards (bio) molecules fixation. Interface Focus. 2014;**4**(1):20130040. DOI: 10.1098/rsfs.2013.0040

[70] Wang S, Wen S, et al. Amino-propyltriethoxysilane-mediated surface functionalization of hydroxyapatite nanoparticles: Synthesis, characterization, and in vitro toxicity assay. International Journal of Nanomedicine. 2011;**6**:3449-3459. DOI: 10.2147/IJN.S27166

[71] Kang SM, Hwang NS, et al. One-step multipurpose surface functionalization by adhesive catecholamine. Advanced Functional Materials. 2012;**22**(14):2949-2955. DOI: 10.1002/adfm.201200177

[72] Lcc H, Dellatore SM, et al. Mussel-inspired surface chemistry for multifunctional coatings. Science (80-). 2007;**318**(5849):426-430. DOI: 10.1126/science.1147241

[73] Zhang K, Wang Y, et al. Bioinspired surface functionalization for improving osteogenesis of electrospun polycaprolactone nanofibers. Langmuir. 2018;**34**(50):15544-15550. DOI: 10.1021/acs.langmuir.8b03357

[74] Ye Q, Zhou F, et al. Bioinspired catecholic chemistry for surface modification. Chemical Society Reviews. 2011;**40**(7):4244-4258. DOI: 10.1039/c1cs15026j

[75] Jin Y, Zhu Z, et al. A facile heparin/carboxymethyl chitosan coating mediated by polydopamine on implants for hemocompatibility and antibacterial properties. Applied Surface Science. 2020;**528**:146539. DOI: 10.1016/j.apsusc.2020.146539

[76] Lee JY, Aguilar LE, et al. UV light assisted coating method of polyphenol Caffeic acid and mediated immobilization of metallic silver particles for antibacterial implant surface modification. Polymers (Basel). 2019;**11**(7):1200. DOI: 10.3390/polym11071200

[77] Tu MG, Lee AKX, et al. Caffeic acid–coated nanolayer on mineral trioxide aggregate potentiates the host immune responses, angiogenesis, and odontogenesis. Journal of Endodontia. 2020;**46**(10):1455-1464. DOI: 10.1016/j.joen.2020.07.003

[78] Song J, Cui N, et al. Sorption studies of tetracycline antibiotics on hydroxyapatite (001) surface—A first-principles insight. Materials (Basel). 2022;**15**(3):797. DOI: 10.3390/ma15030797

[79] He Q, Pan L, et al. Bioinspired synthesis of large-pore, mesoporous hydroxyapatite nanocrystals for the controlled release of large pharmaceutics. Crystal Growth & Design. 2015;**15**(2):723-731. DOI: 10.1021/cg501515c

[80] El Bekkali C, Labrag J, et al. Porous ZnO/hydroxyapatite nanomaterials with effective photocatalytic and antibacterial activities for the degradation of antibiotics. Nanotechnology for Environmental Engineering. 2022;7(2):333-341. DOI: 10.1007/s41204-021-00172-7

[81] Song J, Li X, et al. Zn and Ag doping on hydroxyapatite: Influence on the adhesion strength of high-molecular polymer polycaprolactone. Molecules. 2022;**27**(6):1928. DOI: 10.3390/molecules27061928

[82] Sebastin AXS, Uthirapathy V. In vitro electrochemical behavior of sol-gel derived hydroxyapatite/graphene oxide composite coatings on 316L SS for biomedical applications. ChemistrySelect. 2020;**5**(39):12140-12147. DOI: 10.1002/slct.202003368

[83] Yigit O, Dikici B, et al. Hydrothermal synthesis of nanocrystalline hydroxy-apatite–graphene nanosheet on Ti-6Al-7Nb: Mechanical and in vitro corrosion performance. Journal of Materials Science. Materials in Medicine. 2021;**32**(4):1-14. DOI: 10.1007/s10856-021-06514-w

[84] Chai F, Hornez JC, et al. Antibacterial activation of hydroxyapatite (HA) with controlled porosity by different antibiotics. Biomolecular Engineering. 2007;**24**(5):510-514. DOI: 10.1016/j.bioeng.2007.08.001

[85] Mohd Yusoff MF, Abdul Kadir MR, et al. Dipcoating of poly (ε-caprolactone)/hydroxyapatite composite coating on Ti6Al4V for enhanced corrosion protection. Surface and Coatings Technology. 2014;**245**:102-107. DOI: 10.1016/j.surfcoat.2014.02.048

[86] Iconaru SL, Predoi MV, et al. Dextran-thyme magnesium-doped hydroxyapatite composite antimicrobial coatings. Coatings. 2020;**10**(1):57. DOI: 10.3390/coatings10010057

[87] Tian M, Cai S, et al. Superhydrophilic hydroxyapatite/hydroxypropyltrimethyl ammonium chloride chitosan composite coating for enhancing the antibacterial and corrosion resistance of magnesium alloy. Progress in Organic Coatings. 2022;**165**:106745. DOI: 10.1016/j.porgcoat.2022.106745

[88] Rahmati M, Lyngstadaas SP, et al. Coating doxycycline on titanium-based implants: Two in vivo studies. Bioactive Materials. 2020;**5**(4):787-797. DOI: 10.1016/j.bioactmat.2020.05.007

[89] Safavi MS, Walsh FC, et al. Electrodeposited hydroxyapatite-based biocoatings: Recent Progress and future challenges. Coatings. 2021;**11**(1):110. DOI: 10.3390/coatings11010110

[90] Li B, Xia X, et al. Biological and antibacterial properties of the micro-nanostructured hydroxyapatite/chitosan coating on titanium. Scientific Reports. 2019;**9**(1):14052. DOI: 10.1038/s41598-019-49941-0

[91] Khanmohammadi S, Aghajani H, et al. Vancomycin loaded-mesoporous bioglass/hydroxyapatite/chitosan coatings by electrophoretic deposition. Ceramics International. 2022;**48**(14):20176-20186. DOI: 10.1016/j.ceramint.2022.03.296

[92] Zheng F, Wang S, et al. Characterization and antibacterial activity of amoxicillin-loaded electrospun nano-hydroxyapatite/poly(lactic-co-glycolic acid) composite nanofibers. Biomaterials. 2013;**34**(4):1402-1412. DOI: 10.1016/j.biomaterials.2012.10.071

[93] Kadkhodaie-Elyaderani A, de Lama-Odría MDC, et al. Medicated scaffolds prepared with hydroxyapatite/streptomycin nanoparticles encapsulated into polylactide microfibers. International Journal of Molecular Sciences. 2022;**23**(3):1282. DOI: 10.3390/ijms23031282

[94] Erdem R, Yavuz E, et al. 52 Electrospinning of antibacterial scaffolds composed of poly (L-lactide-co-ε-caprolactone)/collagen type I/silver doped hydroxyapatite particles: Potential material for bone tissue engineering. Journal of the Textile Institute. 2022;**1**:1-15. DOI: 10.1080/00405000.2022.2046305

[95] Taha M, Chai F, et al. Evaluation of sorption capacity of antibiotics and antibacterial properties of a cyclodextrin-polymer functionalized hydroxyapatite-coated titanium hip prosthesis. International Journal of Pharmaceutics. 2014;**477**(1-2):380-389. DOI: 10.1016/j.ijpharm.2014.10.026

[96] Sivaraj D, Vijayalakshmi K. Enhanced antibacterial and corrosion resistance properties of Ag substituted hydroxyapatite/functionalized multiwall carbon nanotube nanocomposite coating on 316L stainless steel for biomedical application. Ultrasonics Sonochemistry. 2019;**59**:104730. DOI: 10.1016/j.ultsonch.2019.104730

[97] Sivaraj D, Vijayalakshmi K, et al. Tailoring Cu substituted hydroxyapatite/functionalized multiwalled carbon nanotube composite coating on 316L SS implant for enhanced corrosion resistance, antibacterial and bioactive properties. International Journal of Pharmaceutics. 2020;**590**:119946. DOI: 10.1016/j.ijpharm.2020.119946

[98] Vyas V, Kaur T, et al. Bio-functionalization of commercially pure titanium with chitosan/hydroxyapatite

biocomposite via silanization: Evaluation of biological performances. Journal of Adhesion Science and Technology. 2017;**31**(16):1768-1781. DOI: 10.1080/01694243.2016.1278070

[99] Dubnika A, Loca D, et al. Functionalized silver doped hydroxyapatite scaffolds for controlled simultaneous silver ion and drug delivery. Ceramics International. 2017;**43**(4):3698-3705. DOI: 10.1016/j.ceramint.2016.11.214

[100] Ahmed S, Alsenany N, Mansour SF, et al. Controlled compositions of tellurium/vanadium co-doped into hydroxyapatite/-polycaprolactone for wound healing applications. New Journal of Chemistry. 2022. DOI: 10.1039/D2NJ00760F

[101] Roohani-Esfahani SI, Nouri-Khorasani S, et al. The influence hydroxyapatite nanoparticle shape and size on the properties of biphasic calcium phosphate scaffolds coated with hydroxyapatite–PCL composites. Biomaterials. 2010;**31**(21):5498-5509. DOI: 10.1016/j.biomaterials.2010.03.058

[102] Al-Wafi R. Polycaprolactone-based antibacterial nanofibrous containing vanadium/hydroxyapatite with morphology, mechanical properties, and in vitro studies. New Journal of Chemistry. 2022. DOI: 10.1039/D1NJ02249K

[103] AlSalem HS, Keshk AA, et al. Physico-chemical and biological responses for hydroxyapatite/ZnO/graphene oxide nanocomposite for biomedical utilization. Materials Chemistry and Physics. 2022;**283**:125988. DOI: 10.1016/j.matchemphys.2022.125988

[104] Mirković M, Filipović S, et al. Hydroxyapatite/TiO_2 nanomaterial with defined microstructural and good antimicrobial properties. Antibiotics. 2022;**11**(5):592. DOI: 10.3390/antibiotics11050592

[105] Afifi M, El-Naggar ME, et al. Compositional adjusting and antibacterial improvement of hydroxyapatite/Nb_2O_5/graphene oxide for medical applications. Journal of Inorganic and Organometallic Polymers and Materials. 2022;**32**(6):2160-2172. DOI: 10.1007/s10904-022-02266-4

[106] Giacomini D, Torricelli P, et al. Monocyclic β-lactams loaded on hydroxyapatite: New biomaterials with enhanced antibacterial activity against resistant strains. Scientific Reports. 2017;**7**(1):2712. DOI: 10.1038/s41598-017-02943-2

[107] Benedini L, Placente D, et al. Adsorption/desorption study of antibiotic and anti-inflammatory drugs onto bioactive hydroxyapatite nano-rods. Materials Science and Engineering: C. 2019;**99**:180-190. DOI: 10.1016/j.msec.2019.01.098

[108] Al-Mofty SED, Karaly AH, et al. Multifunctional hemostatic PVA/chitosan sponges loaded with hydroxyapatite and ciprofloxacin. ACS Omega. 2022;**7**(15):13210-13220. DOI: 10.1021/acsomega.2c00654

[109] Zhang J, Lei X, et al. Effect of antibacterial Enoxacin on the properties of injectable Nano-hydroxyapatite/polyurethane cement for bone repairing. Journal of Bionic Engineering. 2022;**19**:483-496. DOI: 10.1007/s42235-021-00144-2

[110] Verma G, Shetake NG, et al. Covalent immobilization of doxorubicin in glycine functionalized hydroxyapatite nanoparticles for pH-responsive release. New Journal of Chemistry. 2018;**42**(8):6283-6292. DOI: 10.1039/C7NJ04706A

[111] Jin X, Xiong Y, et al. Self-adaptive antibacterial porous implants with sustainable responses for infected bone

defect therapy. Advanced Functional Materials. 2019;**29**(17):1807915. DOI: 10.1002/adfm.201807915

[112] Koizumi D, Suzuki K, et al. Preparation of protamine-adsorbed calcium phosphate powders and their antibacterial property. Journal of Asian Ceramic Societies. 2022;**10**(1):230-240. DOI: 10.1080/21870764.2022.2035488

[113] Chen X, Sevilla P, et al. Surface biofunctionalization by covalent co-immobilization of oligopeptides. Colloids Surfaces B Biointerfaces. 2013;**107**:189-197. DOI: 10.1016/j.colsurfb.2013.02.005

[114] Li K, Chen J, et al. Polymer brush grafted antimicrobial peptide on hydroxyapatite nanorods for highly effective antibacterial performance. Chemical Engineering Journal. 2021;**423**: 130133. DOI: 10.1016/j.cej.2021.130133

[115] Arregui M, Latour F, et al. Ion release from dental implants, prosthetic abutments and crowns under physiological and acidic conditions. Coatings. 2021;**11**(1):1-12

[116] Eliaz N. Corrosion of metallic biomaterials: A review. Materials (Basel, Switzerland). 2019;**12**(3):407. DOI: 10.3390/ma12030407

[117] Manivasagam G, Dhinasekaran D, et al. Biomedical implants: Corrosion and its prevention -a review. Recent Patents on Corrosion Science. 2010;**2**:40-54

[118] Kokubo T, Takadama H. How useful is SBF in predicting in vivo bone bioactivity? Biomaterials. 2006;**27**(15):2907-2915

[119] Franken NAP, Rodermond HM, et al. Clonogenic assay of cells in vitro. Nature Protocols. 2006;**1**(5):2315-2319

[120] Repetto G, del Peso A, et al. Neutral red uptake assay for the estimation of cell viability/cytotoxicity. Nature Protocols. 2008;**3**(7):1125-1131

[121] Plumb JA. Cell sensitivity assays: The MTT assay. In: Langdon SP, editor. Cancer Cell Culture: Methods and Protocols. Totowa, NJ: Humana Press; 2004. pp. 165-169. DOI: 10.1385/1-59259-406-9:165

[122] Rothaeusler K, Baumgarth N. Assessment of cell proliferation by 5-Bromodeoxyuridine (BrdU) labeling for multicolor flow. Cytometry. 2007; Supplement 40:7.31.1-7.31.13. DOI: 10.1002/0471142956.cy0731s40

[123] Duque A, Rakic P. Different effects of Bromodeoxyuridine and [3H]thymidine incorporation into DNA on cell proliferation, position, and fate. The Journal of Neuroscience. 2011;**31**(42):15205-15217. DOI: 10.1523/JNEUROSCI.3092-11.2011

[124] Ali Y, Abd HS. Human Topoisomerase II Alpha as a Prognostic Biomarker in Cancer Chemotherapy. Vol. 37. Netherlands: Tumor Biology. Springer; 2016. pp. 47-55

[125] Mosmann T. Rapid colorimetric assay for cellular growth and survival: Application to proliferation and cytotoxicity assays. Journal of Immunological Methods. 1983;**65**(1):55-63. DOI: 10.1016/0022-1759(83)90303-4

[126] Monteiro-Riviere NA, Inman AO, et al. Limitations and relative utility of screening assays to assess engineered nanoparticle toxicity in a human cell line. Toxicology and Applied Pharmacology. 2009;**234**(2):222-235. DOI: 10.1016/j.taap.2008.09.030

[127] Weinstein M. M02 Performance Standards for Antimicrobial Disk Susceptibility Tests. 13th ed. Clinical and Laboratory Standards Institute. Berwyn, PA; 2018

[128] Nijs A, Cartuyvels R, et al. Comparison and evaluation of Osiris and

Sirscan 2000 antimicrobial susceptibility Systems in the Clinical Microbiology Laboratory. Journal of Clinical Microbiology. 2003;**41**(8):3627-3630. DOI: 10.1128/JCM.41.8.3627-3630.2003

[129] Clancy CJ, Huang H, et al. Characterizing the effects of Caspofungin on Candida albicans , Candida parapsilosis, and Candida glabrata isolates by simultaneous time-kill and Postantifungal-effect experiments. Antimicrobial Agents and Chemotherapy. 2006;**50**(7):2569-2572. DOI: 10.1128/AAC.00291-06

[130] Alicia GL, Amel A, et al. Analysis of the influence of tween concentration, inoculum size, assay medium, and Reading time on susceptibility testing of aspergillus spp. Journal of Clinical Microbiology. 2005;**43**(3):1251-1255. DOI: 10.1128/JCM.43.3.1251-1255.2005

Chapter 6

Use of Plasma Pseudocholinesterase as a Predictor of Mortality in Organophosphate Poisoning

Siva Kumar V., Shruthi K. Siva Kumar, Ipsita Debata, Tejas J. and Viswanathan K. Gowda

Abstract

The study was conducted on patients of organophosphate poisoning admitted to Bapuji Hospital (J. J. M. Medical College), Davangere during a period of October 2011 to March 2013. To know the incidence of acute Organophosphate poisoning, epidemiological aspects of the patient and plasma pseudocholinesterase levels at the time of admission and correlation within hospital mortality. Total number of cases studied were 150. At the time of admission blood was drawn for estimation of plasma pseudocholinesterase estimation. The patients were clinically divided into three grades according to Dreishbachs criteria. Analysis was performed by cobas integra 400 cholinesterase assay system. All patients were followed-up for 3 days to know the outcome. Majority of the cases (40%) belong to 21 to 30 years age group and predominantly belonged to male sex (73%). Seventy eight cases (52%) had severe poisoning, 40 cases (26.67%) had moderate poisoning and 32 cases (21.33%) had mild poisoning. Sixty cases (40%) had fatal outcome. Suicidal consumption was seen in 128 cases (85.33%). Plasma pseudocholinesterase levels associated with fatalities in severe poisoning and was found to range from 912 to 2490 U/L which accounts to suppression of plasma pseudocholinesterase levels by 84.04 to 93.19%.

Keywords: organophosphate poisoning, plasma pseudocholinesterase, butyrlycholinesterase, suicidal poisoning, agricultural poisons

1. Introduction

Poison is a substance (solid, liquid or gaseous), which if introduced in the living body or brought into contact with any part thereof, will produce ill-health or death by its constitutional or local effects or both. However, Goethe says that, 'There is no such thing as poison, it all depends on dose'.

It might be challenging to draw a line between a medicine and a poison because a medicine can behave as a poison in big amounts and can be a medicine in tiny doses. The "intent" with which they are intentionally supplied, as opposed to accidently, is the sole significant distinction.

Many people believe that toxicology, or thc study of poisons, is a very young field of science. On the other hand, evidence of the damaging effects of chemicals on living things dates back to prehistory. Even in the past, mankind looked for poison antidotes. While many chemical compounds used to make medications can behave as poisons in their larger abundance, there has been an almost centuries-long risk to both human health and the environment.

In all civilised nations, poisoning incidents are steadily rising. Depending on a number of variables, the type of poison employed for distinct modalities may change. However, there has been a steady increase in accidental and suicidal poisoning in agriculture and domestic settings. The increased usage of many chemical products in the home is blamed for the increase in accidental poisoning among youngsters. Children are most likely to become poisoned in the kitchen (34%), bedroom (27%), bathroom, and laundry rooms (15%). Among children the common poisons include kerosene, household chemicals, drugs, pesticides and garden plants. Industrial poisoning is gradually receding, owing to advances in industrial hygiene and medical service and to the increasing automation of industrial processes.

In adults the manner of poisoning, irrespective of the sex can be;

1. Suicidal
2. Accidental
3. Homicidal
4. Self-treatment
5. Injudicious medication.

Poisoning can happen as a result of:

1. The use of poison for illegal objectives.
2. Ingesting poison by accident while thinking it is a harmless material.
3. Accidental or unintentional inhalation of poisonous gas.
4. Improperly mixing poison-containing medications.
5. Accidentally taking a huge dosage of medicine that is poisonous.
6. Abundant self-medication.
7. Drug addiction.
8. Being bit by a dangerous animal.
9. Food contaminated with poisons or microbes.

Criteria for an ideal suicidal Poison:

An ideal suicidal poison should be:

a. Easily available.

b. Cheap.

c. Tasteless, if not, have a pleasant taste.

d. Highly toxic and sure in action.

e. Capable of being easily consumed with food or drink.

f. Capable of producing painless death, preferably through sleep.

Opium and barbiturates satisfy several of the above criteria. But organ phosphorus compounds and endrin commonly used for the purpose. The substances like oleander seeds, oxalic acid, carbolic acid, aspirin, arsenic trioxide, mercuric chloride, or coal-gas inhalation may be used for the purpose.

Criteria for an ideal homicidal Poison:

An ideal homicidal poison should be:

a. Colourless, tasteless and odourless.

b. Capable of being easily administered in food, drink, or medicine without arousing any suspicion.

c. Should be highly toxic and certain in its effects.

d. Signs and symptoms of it should resemble a natural disease without raising any suspicion.

e. Signs and symptoms are to appear late, giving sufficient time to the culprit to escape or to avoid suspicion.

f. Having no good antidote against the poison.

g. Having no specific postmortem findings to arouse suspicion.

h. To be rendered undetectable from the body by toxicological examination.

Organic compounds of "fluorine" used as rodenticides and "thallium" satisfy several of the above criteria. However, compounds of arsenic, aconite, antimony, mercury, copper, powdered glass, oleander, nuxvomica, madar etc. may be used for the purpose of homicide.

Apart from poison those are ingested, poisoning due to animal bites especially snake bites are quite common in India. Except in Arctic lands, New Zealand and Ireland snakes are found all over the world. In India, snake bites are usually accidental in nature. Especially in southern districts of West Bengal, Orissa, Assam, Bihar, Madhya Pradesh, Karnataka, Andhra Pradesh etc., the incidence is high. At least more than 20,000 persons die per year out of 2 lakh snake bite cases in India.

Human poisoning due to suicide, homicidal, accidental is common in India, as poisons are easily obtained in the market such as insecticides, pesticides, rodenticides, weed killers, and drugs. In addition to the above, many poisonous plants grow widely all over the country are also used in poisoning e.g., Oleander, Aconite, Nux- vomica, Calotropis, Datura, Nerium odorum, Abrus precatorius etc. Many Indians consider taking off life by poisoning a lesser crime than bloodshed Reddy et al. [1]. Incidence of accidental poisoning is also increasing because of increasing use of chemicals both for industrial and also domestic purposes. Insecticides and weed killers are also in extensive use for agricultural purposes.

Following the knowledge of the highly lethal nature of these substances they have become popular as suicidal and homicidal poisons. In Belgaum, the age-old tradition of suicides by drowning in wells or by hanging have been replaced by poisoning oneself by the use of organophosphorus compounds, etc.

1.1 Objectives

- Incidence of acute Organophosphate poisoning.
- Epidemiological aspects of the patient.
- Pseudocholinesterase levels at the time of admission and correlation within hospital mortality.

Organophosphates are a group of compounds with various toxicities to different form of life. The widest use of these compounds is as insecticides.

Organophosphate insecticides have controlled vectors of Malaria. Their use is increasing since, low toxic organophosphates are now replacing Chlorinated hydrocarbon insecticides such as DDT, which accumulates unchanged in human and animal tissues and have adverse effects.

The organophosphate insecticides are esters and oxides of phosphoric and pyrophosphoric acid which, when introduced into animal body, inhibit the enzymes that hydrolyse acetylcholine. They are called anticholinesterase agents. These inhibitors have frequently been called "irreversible" inhibitors because it was believed that the enzyme attacked by them is permanently destroyed and that recovery took place by formation of new enzyme molecules.

In India, the first report of oral poisoning by Organophosphorus compounds was reported by Mutalik et al. [2]. They studied 25 cases of Diazinon poisoning and described the various clinical features, management and autopsy findings.

Acetyl choline (Ach) an ester of choline is present in various organs and tissues of the body. It plays an important role in transmission of nerve impulses at - Synapses & Myoneuronal junction. Acetylcholine is rapidly destroyed by an enzyme Acetyl choline esterase (AchE). This enzyme stops the action of acetylcholine which is present in various body tissues; including muscles, nerve cells and red blood cells. A deficiency of cholinesterase results in neuromuscular excitability, a prominent clinical feature in.

1.2 Organophosphate poisoning

According to Dr. K.S.N. Reddy [1], the most commonly used poison in rural and urban places in South India is organophosphorus compounds, which are powerful inhibitors of cholinesterase enzyme. Inactivating it by phosphorylation at myoneuronal junction, it results in a syndrome of over activity due to excess of unhydrolysed acetyl choline at myoneuronal junction, which leads to accumulation of Acetylcholine

Concentration of AchE	Severity of toxicity
20–25%	Mild
10–20%	Moderate
<10%	Severe

Table 1.
Callaway classification based on acetylcholine esterase suppression.

at parasympathetic, sympathetic and somatic sites. Thus preventing the transfer of nerve impulses across the myoneuronal junction.

According to Callaway et al. [3], the red cell choline esterase level in good health ranges between 75 and 142 units. Mild symptoms occur when acetylcholine esterase activity reduces to 20–25% of normal. If moderate poisoning occurs, the activity of AchE decreases to 10–20% of normal. Severe poisoning results in an activity of less than 10% of normal (**Table 1**).

This clearly indicates that the rate limiting factor in a case of organophosphorus compound poisoning is the concentration of acetyl choline esterase enzyme at myoneuronal junction.

Thus the concentration of AchE at myoneuronal junction acts as a guide to determine - (i) The severity of toxicity, (ii) The therapeutic dose of atropine and (iii) PAM (Pyridine Aldoxime ethiodide), so that these antidotes, may not be used in excess quantity than required, because they themselves are capable of causing harmful effects on the body.

According to the text book of Modern Toxicology by V.V. Pillay [4], Plasma cholinesterase levels (Pseudocholinesterase) are diagnostic.

In the present study the concentration of AchE was estimated in the plasma in order to assess the severity and mortality.

Clinical Manifestations of Organophosphorus Poisoning:

Organophosphorus compound produces clinical manifestations by depression of the enzymes cholinesterase, resulting in the accumulation of acetylcholine at various receptors. This has three types of effect.

1. Cholinomimetic actions of muscarinic type at autonomic effector organs.

2. Nicotinic actions: Stimulations of all autonomic ganglion and skeletal muscle.

3. CNS effect; Stimulation with consequent depression of cholinoceptive sites in the CNS [5–7].

1.3 Classification of organophosphate poisoning based on clinical features

The severity of poisoning is graded according to modified version of Dreisbachs classification (**Table 2**) [8].

Plasma cholinesterase activity recovers slowly due to the irreversible nature of organophosphate inhibition. Without the use of pralidoxime, plasma cholinesterase rises an average of 15.6% over 14 days in one group of organophosphate-exposed workers. The serial levels rather than one initial level may be valuable in diagnosing organophosphorus poisoning. The poor correlation between acetylcholinesterase level and clinical effects may mislead clinicians, into making incorrect diagnosis of mild

Grade (Dreisbachs)	Symptom
Mild	1. Nausea 2. Vomiting 3. Diarrhoea 4. Sweating
Moderate	1. Lacrimation 2. Salivation 3. Miosis 4. Fasciculation
Severe	1. Coma 2. Seizures 3. Incontinence 4. ARDS 5. Areflexia

Table 2.
Dreisbachs criteria.

poisoning. Sequential post-exposure determinations may be necessary to confirm Acetylcholinesterase inhibition. Initially acetylcholinesterase should regenerate by 15 to 20% within 3 to 5 days [9, 10].

1.4 Anti-cholinesterase agents

These compounds are capable of inhibiting cholinesterase enzyme both true and pseudo and thus resulting in accumulation of acetylcholine at various cholinergic sites. Thus, pharmacological effects resulting from administration of anticholinesterase resemble the actions of endogenous acetylcholine or exogenously administered acetylcholine.

Classification:

1. Reversible anti-cholinesterase - physostigmine, neostigmine.

2. Irreversible anti-cholinesterase - organophosphorus compounds.

1.5 The reversible anti-cholinesterase

Reversible anticholinesterase by their structural resemblance to acetylcholine are capable of combining with anionic and esteratic sites of cholinesterase as well as with acetylcholine receptors. However, the complex which they form with the esteratic site of cholinesterase is much less readily hydrolysed from acetyl esteratic site than the complex formed with acetylcholine. This produces a temporary inhibition of enzyme.

Reversible anticholinesterase have gained therapeutic importance. They are found to be beneficial in the treatment of glaucoma, myasthenia gravis, paralytic ileus, urinary retention, in the treatment of certain cardiac arrhythmias (paraoxysmal supra ventricular tachycardia) and in Belladona poisoning.

1.6 The irreversible anti-cholinesterase

The irreversible anticholinesterase combines only with the esteratic site ofcholinesterase which is phosphorylated. The hydrolysis of phosphorylated site is extremely slow and in certain cases does not occur at all this results in irreversible inhibition of enzyme. The irreversible anticholinesterase are of limited therapeutic use because of prolonged action and high toxicity. They are found to be of great benefit in the field of agriculture, where they are used as insecticides.

1.7 Paraoxonase

Paraoxonases are a group of enzymes involved in the hydrolysis of organophosphates parathion, diazinon and chlorpyrifos [11, 12]. The important discoveries that certain organophosphorus (OP) insecticides could be enzymatically hydrolyzed by plasma and that this reaction is catalysed by enzymes which were named "A-esterases", were reported in the 1940s and 1950s. There are 3 proteins in this family which include PON-1, PON-2 and PON-3 the genes of which are located in long arm of chromosome 7 [13].

1.8 Chemical structure with classification

Organophosphorus insecticides are usually esters, amides or thiol derivatives of phosphoric or phosphonic acid. The general formula being described in **Figure 1**.

Typically, Rl and R2 are straightforward alkyl or aryl groups. The "laving group," often known as Group "X," can be any of a wide range of substituted or branched aliphatic, aromatic, or heterocyclic groups that are connected to phosphorus by a bond with some degree of liability. Typically -O- or -S-. The similar chemical is known as a phosphate or phosphorothioate, and the double bond may be O or S. Many pesticides are produced in the intrinsically more stable P=S form, which can later be transformed in vivo to the physiologically active oxon [14].

Phosphorothioate oxidation to phosphates poses a risk since phosphates are more volatile and may undergo directly harmful oxidation at higher temperatures. Some formulations (such as malathion) may become more harmful when they are isomerized while being stored in warm, humid circumstances [15].

2. Methodology

The study was conducted on patients of organophosphate poisoning admitted to tertiary care hospital at Davangere, Karnataka, India during the period between October 2011 & March 2013.

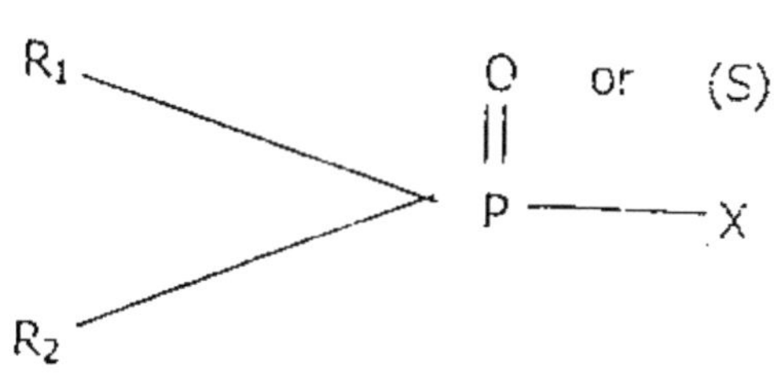

Figure 1.
Chemical structure of Organophosphate.

Materials used for the study were 18-gauge needle, 10 cc syringe, Colour coded vacutainer tubes, Cobas Integra 400 cholinesterase assay system.

Total number of cases studied were 150. At the time of admission blood was drawn for estimation of plasma pseudocholinesterase estimation. A laboratory reference range of 8000 to 18,000 U/L was used in this study which coincide with the estimated population mean levels.

Information was gathered from patient case histories, hospital MLC files, eyewitness interviews, family and friends of the dead, and the investigating officer.

The study covered patients of either sex who had Organophosphate poisoning and were older than 14 years.

All patients with age less than 14 years, poisoning other than organophosphate were excluded from the study.

From the patient, his or her relatives, and the police, a thorough history was gathered on the quantity of poison, the type of poison, etc. Whenever possible, the poison container was also examined. Each patient underwent a clinical examination, and using Dreishbach's criteria, the patients were graded into mild, moderate, and severe groups based on their signs and symptoms. All patients were followed-up for 3 days to know the outcome. In fatal cases the Forensic Science Laboratory was also used for conformation of organophosphorus poisoning. Data was wrangled using Microsoft excel 2016 and analysed with IBM SPSS v26.

3. Results and discussion

Out of 150 cases, highest number of poisonings was reported in the age group of 21 to 30 years (60 patients, 40%) as depicted in **Figure 2**. This correlates with the age which was also reported by S Singh et al., were the mean age of the patient was found to be 26.44 years [16].

Males were found to be more affected (73%) than the females (27%). Similar observations were made by Singh et al. [16], consisting of 67.95% males as depicted in **Figure 3**.

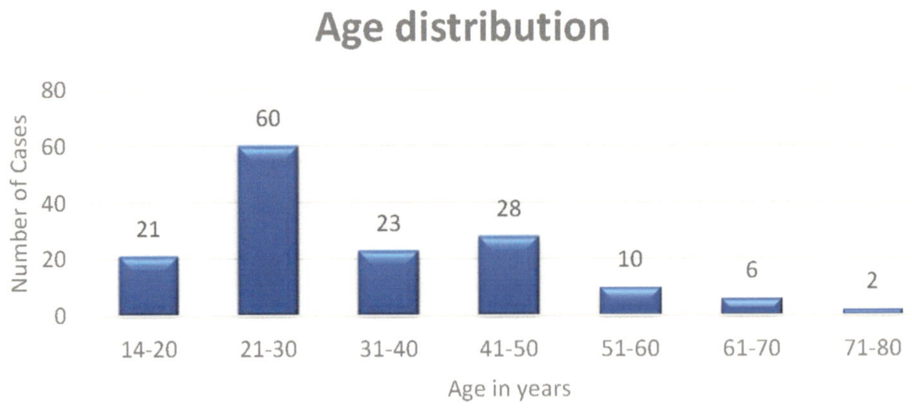

Figure 2.
Age distribution.

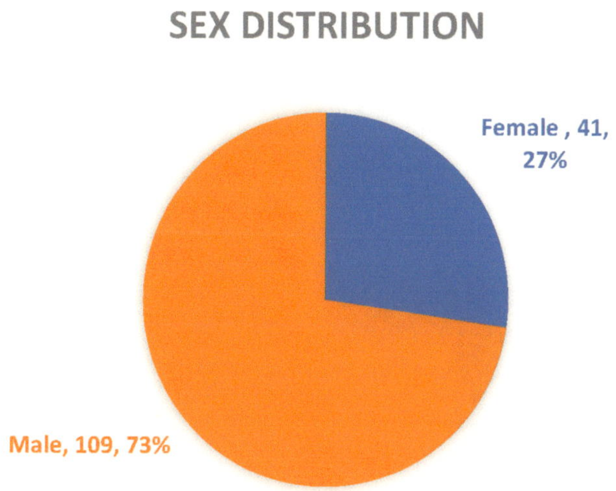

Figure 3.
Sex distribution.

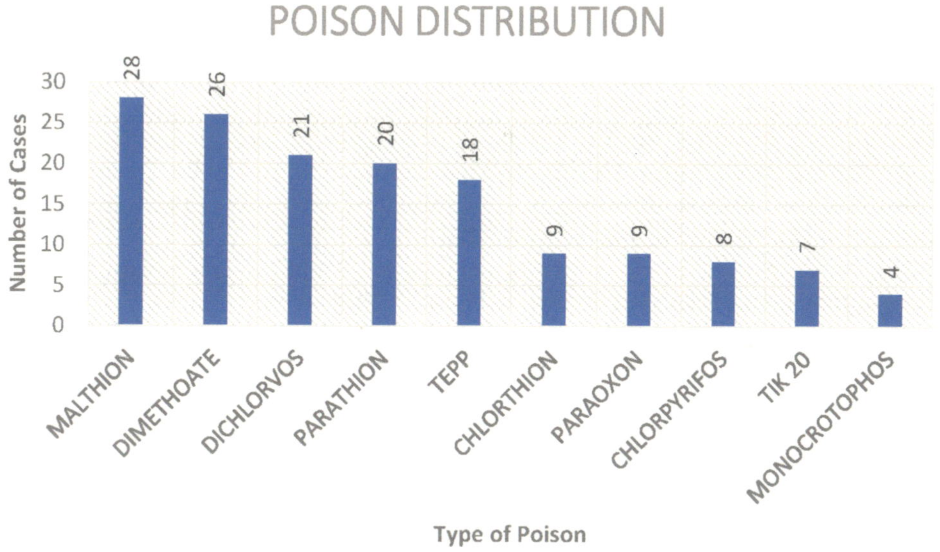

Figure 4.
Type of poison distribution.

3.1 Time taken to reach hospital

The average time taken by the patients to reach the hospital in this study was found to be 5 hours with a minimum of 1 hour and a maximum of 12 hours. This can be attributed to the inefficient health care delivery and transport systems and wide area of coverage leading to lack of access to health care services to most of the areas situated in the outskirt of the city.

The commonest poison consumed in the study was Malathion, 28 patients (18.67%) as shown in **Figure 4**. Second common was Dimethoate, 26 patients (17.33%). In every

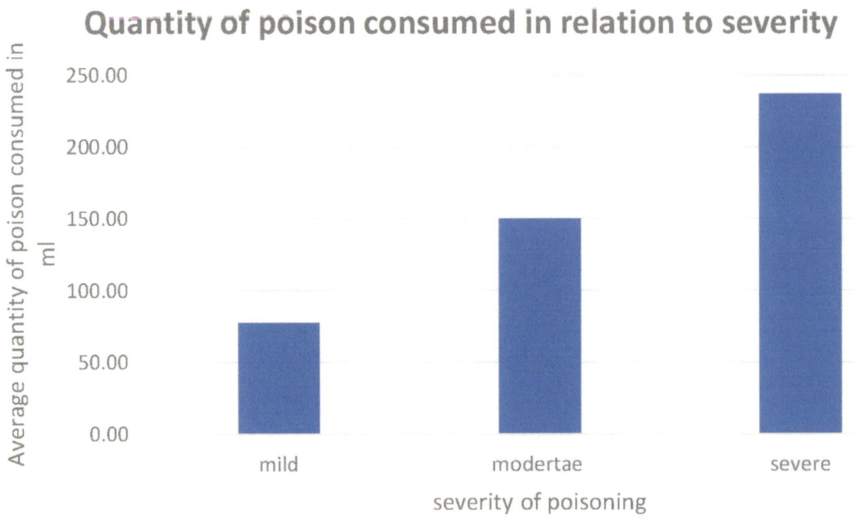

Figure 5.
Approximate quantity of insecticide ingested.

instance, the poisons were taken orally. One of the most widely used organophosphate insecticides is malathion, which is frequently accessible for agricultural application. Due to its simple accessibility to farmers and fatal action, despite its unpleasant taste, it is most frequently used orally. Other research by Namba, Greenfield and Grob [6]; Daglia & Shaikh [17]; Wadia, Bhirud, Gulavani & Amin [18] and Wille, Thiermann & Worek [19] also reflect similar finding. An acute case of demeton poisoning in a child was reported by Felsenstein and collegues [20].

Most of the participants in our study had ingested 101 to 200 ml of an organophosphorus chemical (43.33%) as seen in **Figure 5**. This amount exceeds the lethal dose for Malathion and Dimethoate, the two most prevalent poisons employed in our investigation.

In relation to severity of poisoning, the average quantity of poison consumed in case of severe poisoning was found to be 237.05 ml which could also be a causative factor in highest mortality being associated with this group as this dose in in excess of the fatal dose of most of the poisons observed in this study.

The patients were clinically examined and divided into groups using the Dreisbach's criteria 78 patients (52%) had sever poisoning, 40 patients (26.67%) had moderate poisoning and 32 patients (21.33%) had mild poisoning (see **Figure 6**). The majority of the patients had serious poisoning. Our may be partially due to the way of death, as the majority of cases in this study drank poison with the intention of killing themselves, in which case the amount consumed would have been substantially more than accidental or extremely infrequently homicidal intake.

During the 3 days that the patients were monitored, 40% of them died (including those who went into a coma), while 60% recovered with the help of treatment. 17.33% of all fatalities—or deaths—occurred within 24 hours of hospital admission (see **Figure 7**). Only 12% of patients with severe poisoning survived, and all deaths were related to it. The mortality rate for patients with suicidal organ phosphorus poisoning was reported to be 26% in a study by Kar [21], which is greater than the rate we discovered because our investigation also looked at other causes of death.

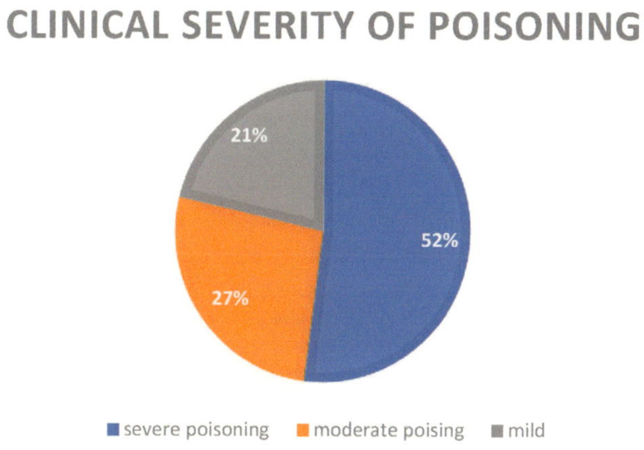

Figure 6.
Clinical severity of poisoning.

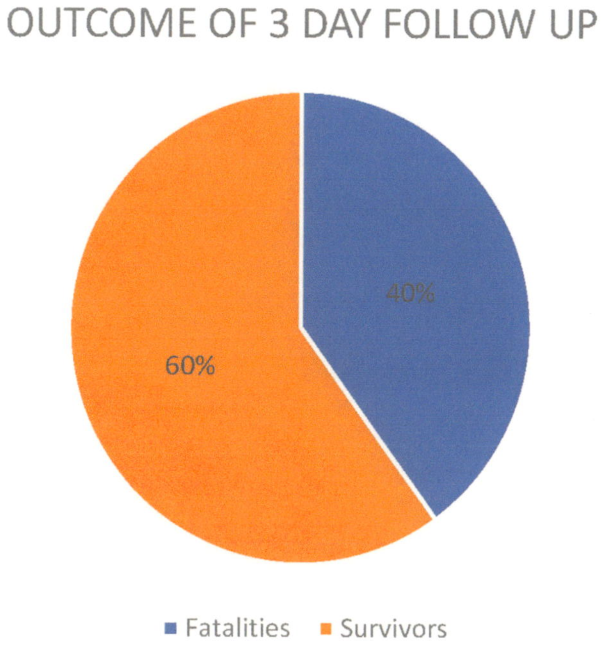

Figure 7.
Final outcome of 3 day follow-up post poisoning.

The mortality rate of 17.30 percent reported in a research by Singh et al. [16] is consistent with our 24-hour mortality rate.

Pseudocholinesterase levels in the blood (**Table 3**; **Figures 8–10**).

Plasma pseudo-cholinesterase levels, as shown in **Table 3** were compared and in the group that had experienced severe poisoning were observed to range from 912 to 2490 U/L (mean value = 1696.62 U/L and S.D. = +/− 438.99 U/L). All 60 fatalities (40%) noted in this study were included in this category. When the plasma levels of each of the 60 fatal instances were compared, it was shown that they were statistically

Clinical grade of poisoning		Plasma Pseudo-Cholinesterase Levels			Percentage suppression of Plasma Pseudocholinesterase level
		Range (U/L)	**Mean (U/L)**	**Standard deviation (U/L)**	
Severe	Death within 24 hours (17.33% mortality rate)	912 to 1678	1380	200.68	89.24 to 93.19%
	Death over a period of 3 days following admission (40% mortality rate)	912 to 2490	2158.15	774.15	84.04 to 93.19%
Moderate		4128 to 7642	5339.40	1121.33	51.01 to 69.19%
Mild		7654 to 11,230	9110.38	927.29	28.01 to 42.88%

Table 3.
Clinical grade of poisoning and its relation to plasma pseudo-cholinesterase.

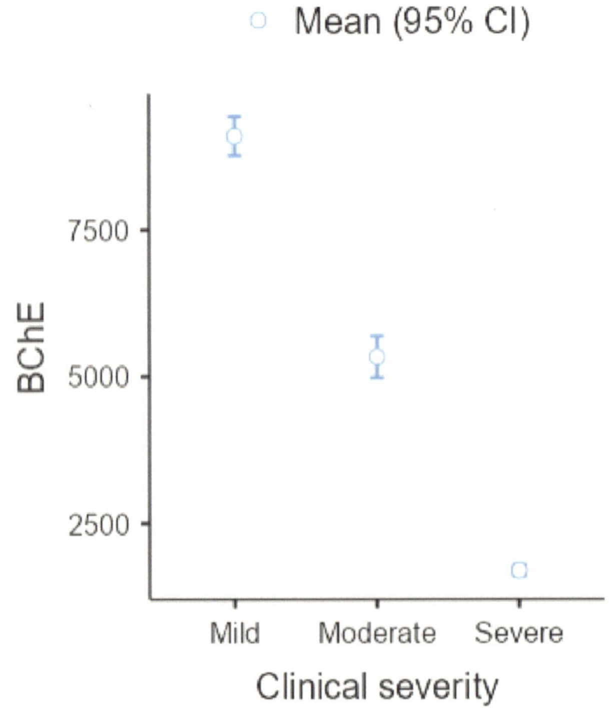

Figure 8.
Clinical severity.

very significant ($p < 0.01$). As a result, plasma pseudo-cholinesterase is suppressed by 84.04 to 93.19%. Even though 18 patients (12%) who had severe poisoning survived, there was no statistically significant difference between them and the fatalities

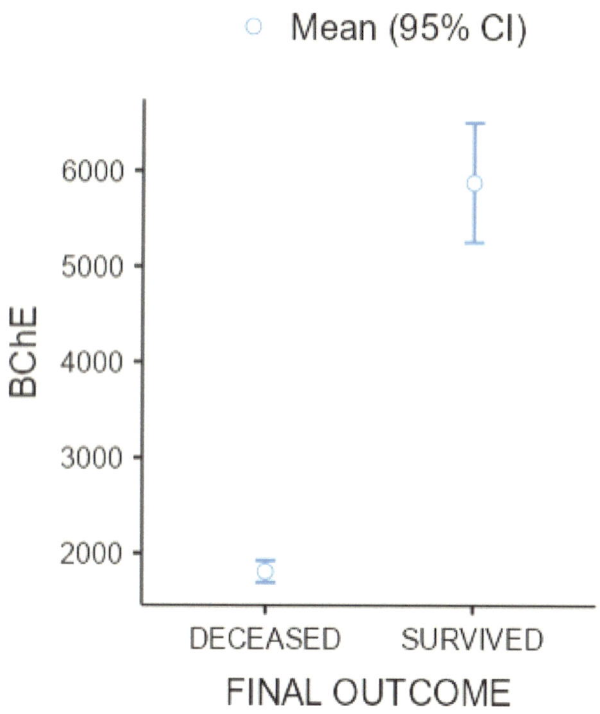

Figure 9.
Final outcome.

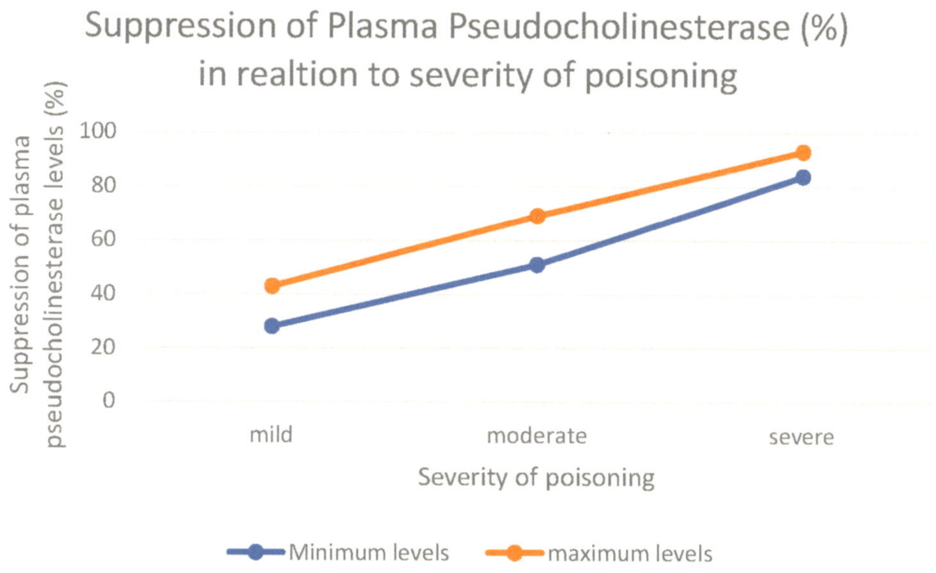

Figure 10.
Suppression of plasma pseudocholinesterase (%) in relation to severity of poisoning.

($p > 0.5$). The clinical severity also worsened with reduction in plasma pseudocholinesterase levels as depicted in **Figure 8** with correspondingly higher fatality as depicted in **Figure 9**.

The plasma-pseudocholinesterase levels were found to be statistically highly significant ($p < 0.01$) and ranged from 912 to 1678 U/L (mean value = 1390.35 U/L and S.D. = +/− 200.68 U/L), suppressing plasma pseudo-cholinesterase by 89.24 to 93.19% in the 26 patients (17.33%) who passed away within 24 hours of hospital admission.

When plasma pseudo-cholinesterase levels in the group with mild poisoning were evaluated, it was discovered that they ranged from 4128 U/L to 7642 U/L (mean value = 5339.40 U/L and S.D. = +/− 1121.33 U/L) and were statistically highly significant ($p < 0.01$). This results in a 51.01 to 69.19 percent reduction in plasma pseudo-cholinesterase.

When plasma pseudo-cholinesterase levels in the group with mild poisoning were analysed, it was discovered that they ranged from 7654 to 11,230 U/L (mean value = 9110.38 U/L and S.D. = +/− 927.29), and that this difference was statistically highly significant (p 0.001). As a result, plasma pseudo-cholinesterase is reduced by 28.01 to 42.88%.

The range of suppression of plasma pseudocholinesterase showed a linear correlation with clinical severity of poisoning as depicted in **Figure 10**.

Severe poisoning accounted for all the 60 fatalities (40%) observed in this study and only 18 patients (12%) survived. The Plasma Pseudocholinesterase levels of all the 60 fatal cases were compared with those who survived and found to be statistically highly significant ($p < 0.01$). As a result, the estimated levels of pseudocholinesterase at the time of hospital admission serve as an excellent prognostic indicator and also aid in the dose adjustment of numerous medicines used for therapy. Inhibition of plasma pseudocholinesterase from 84.04 to 93.19% was linked to severe poisoning when fatality occurred well over 3 days after hospital admission. Plasma pseudocholinesterase inhibition of 89.24 to 93.19% (as observed in patients who passed away within 24 hours of arrival) is linked with 100% mortality. This demonstrates a connection between plasma pseudocholinesterase and poisoning severity. And a fatal outcome is linked to the inhibition of this enzyme by more than 89.24% (i.e., plasma pseudocholinesterase levels below 1678 U/L). This is consistent with a research by Xu et al. [22] that found severe acute organophosphorus poisoning happens when plasma pseudocholinesterase levels approach 10%. This study's findings concur with a cohort study by Eddleston and collegues [23] who discovered that plasma pseudocholinesterase activity of 600 U/L at admission was highly specific for dimethoate toxicity and very sensitive to chlorpyrifos poisoning. According to Sunder Ram et al. [8] plasma pseudocholinesterase levels below 10% of normal are associated with a poor prognosis, which is consistent with the findings of this investigation. In acceptance with this study are studies by Reddy [1], Pillay [4] and Sozmen and collegues [24] as well. No discernible difference was seen in the levels of pseudocholinesterase in postmortem samples from brought-dead subjects and partially treated cases, according to Kukde and collegues [25].

Plasma pseudocholinesterase concentrations in mild and moderate poisoning were also found to be statistically highly significant ($p < 0.01$) and hence the observed levels can be effectively used in assessing the patient outcome and also for calibration of dose of pralidoxime which is the specific antidote.

4. Conclusion

1. Total of 150 cases were included in the present study who had organophosphate poisoning.

2. Majority of the cases belonged to age group between 21 to 30 years and the incidence among males (73%) is higher than that in females (27%).

3. All the cases were agricultural farmers by occupation.

4. Average time taken by the cases to reach hospital was 5 hours.

5. Malathion, Dimethoate and Dichlorvos were the most common poisons consumed in decreasing order.

6. Majority of the cases belonged to severe grade (52%) of poisoning presenting with coma and convulsions.

7. Malathion and parathion were associated with highest mortality (15% each).

8. Major patients (43.33%) consumed between 101 to 200 ml of the poison and most of them were in the severe group with high mortality.

9. There is a fairly good correlation between clinical severity of organophosphorus poisoning and Plasma Pseudocholinesterase levels and the plasma pseudocholinesterase levels equal to or less than 1390.35 U/L (+/− 200.68 U/L) was fatal. This amounts to suppression of plasma pseudocholinesterase by 90.414% (+/−1.384%).

Acknowledgements

I sincerely thank Dr. Bandenawaz my beloved senior and now currently Professor and Head of FMT in Navodaya Medical College Hospital & Research Center, Raichur for his constant help and encouragement during my study. His words of wisdom and timely help and support have helped me a great deal in accomplishing this study.

Conflict of interest

Nil

Notes/thanks/other declarations

Nil

Author details

Siva Kumar V.[1], Shruthi K. Siva Kumar[2], Ipsita Debata[3], Tejas J.[4*]
and Viswanathan K. Gowda[5]

1 Department of Forensic Medicine and Toxicology, Apollo Institute of Medical Sciences and Research, Chittor, India

2 Department of Physiology, St. Peter's Medical College, Hospital and Research Institute, Hosur, India

3 Department of Community Medicine, Akash Institute of Medical Sciences and Research Centre, Bengaluru, India

4 Department of Forensic Medicine and Toxicology, Karpaga Vinayaga Institute of Medical Sciences and Research Center, Chengalpattu, India

5 Department of Forensic Medicine and Toxicology, Basaveshwara Medical College and Hospital, Chitradurga, India

*Address all correspondence to: tejas05in@gmail.com

References

[1] Reddy KN. The Essentials of Forensic Medicine and Toxicology. New Delhi, India: Jaypee Brothers Medical Publishers; 2016. pp. 516-523

[2] Mutalik GS, Wadia RS, Pai VR. Poisoning by diazinon, an organophosphorous insecticide. Journal of the Indian Medical Association. 1962;**38**:67-71

[3] Callaway S, Davies D, Rutland J. Blood cholinesterase levels and range of personal variation in a healthy adult population. British Medical Journal. 1951;**1949**:812-816

[4] Pillay V. Modern Medical Toxicology. New Delhi: Jaypee Brothers Medical Publishers; 2013. pp. 386-410

[5] Taylor P. Anticholinesterase agents. In: Brunton LL, Chabner BA, Knollmann BC, editors. Goodman & Gilman's: The Pharmacological Basis of Therapeutics. 12th ed. New York, NY: McGraw-Hill Education; 2015 [cited 2022 Jun 8]. Available from: accessmedicine.mhmedical.com/content.aspx?aid=1127865480

[6] Namba T, Greenfield M, Grob D. Malathion poisoning. A fatal case with cardiac manifestations. Archives of Environmental Health. 1970;**21**(4):533-541

[7] Crossland J, Malone MH. Lewis's pharmacology. In: Journal of Pharmaceutical Sciences. 4th ed. Vol. 60. Philadelphia, USA: Williams and Wilkins; 1971

[8] Ram JS, Kumar SS, Jayarajan A, Kuppuswamy G. Continuous infusion of high doses of atropine in the management of organophosphorus compound poisoning. The Journal of the Association of Physicians of India. 1991;**39**(2):190-193

[9] Joshi S, Singh M, Deshpande A. Organophsophorous Poisoning - Clinical Profile with Special Reference to Pancreatic and Neurological Involvement. J Sir JJ Group Hosp Grant Med Coll. Vol. 35;1992:3.

[10] Moore PG, James OF. Acute pancreatitis induced by acute organophosphate poisoning. Postgraduate Medical Journal. 1981;**57**(672):660-662

[11] Aldridge WN, Serum esterases. I. Two types of esterase (A and B) hydrolysing p-nitrophenyl acetate, propionate and butyrate, and a method for their determination. The Biochemical Journal. 1953;**53**(1):110-117

[12] Mazur A. An enzyme in animal tissues capable of hydrolyzing the phosphorus-fluorine bond of alkyl fluorophosphates. The Journal of Biological Chemistry. Jul 1946;**164**:271-289

[13] Costa LG, Giordano G, Cole TB, Marsillach J, Furlong CE. Paraoxonase 1 (PON1) as a genetic determinant of susceptibility to organophosphate toxicity. Toxicology. 2013;**307**:115-122

[14] Karalliedde L, Senanayake N. Organophosphorus insecticide poisoning. British Journal of Anaesthesia. 1989;**63**(6):736-750

[15] Macholz R. Organophosphorus Insecticides: A General Introduction (Environmental Health Criteria 63). 181 Seiten. World Health Organization, Geneva 1986. Food Nahrung. 1988;**32**(7):708

[16] Singh S, Sharma BK, Wahi PL, Anand BS, Chugh KS. Spectrum of acute poisoning in adults (10 year experience). The Journal of the Association of Physicians of India. 1984;**32**(7):561-563

[17] Dagli AJ, Shaikh WA. Pancreatic involvement in malathion--anticholinesterase insecticide intoxication. A study of 75 cases. British Journal of Clinical Practice;**37**(7-8):270-272

[18] Wadia RS, Bhirud RH, Gulavani AV, Amin RB. Neurological manifestations of three organophosphate poisons. The Indian Journal of Medical Research. 1977;**66**(3):460-468

[19] Wille T, Thiermann H, Worek F. In vitro kinetic interactions of DEET, pyridostigmine and organophosphorus pesticides with human cholinesterases. Chemico-Biological Interactions. 2011;**190**(2-3):79-83

[20] Felsenstein WC, Staiff DC, Miller GC. Acute demeton poisoning in a child. Archives of Environmental Health. 1976;**31**(5):266-269

[21] Kar N. Lethality of suicidal organophosphorus poisoning in an Indian population: Exploring preventability. Annals of General Psychiatry. 2006;**5**:17

[22] Xu C, Zhang XG, Yang X, He YZ. The diagnostic value of butyrylcholinesterase in acute organophosphorus pesticide poisoning. Zhongguo Wei Zhong Bing Ji Jiu Yi Xue. 2010;**22**(4):193-196

[23] Eddleston M, Eyer P, Worek F, Sheriff MH, Buckley NA. Predicting outcome using butyrylcholinesterase activity in organophosphorus pesticide self-poisoning. QJM. 2008;**101**(6):467-474

[24] Sozmen B, Peker S, Kaya U, Erkan M, Sozmen EY. Markers of long-term exposure to organophosphorus pesticides in farmers who work in viniculture and tobacco production in Turkey. Toxicology Mechanisms and Methods. 2007;**17**(7):379-384

[25] Kukede HG, Ambade VN, Batra AK, Keoliaya AN. Significance of serum cholinesterase level in organophosphate poisoning. Medico-Leg Update. 2012;**12**(2):70-74

Chapter 7

Perspective Chapter: Optical Remote Sensing for Fluorapatite Content Estimation

Nouha Mezned

Abstract

Remote sensing techniques are an interesting alternative to traditional methods for the rapid abundance prediction and mapping of phosphate mineralization surface states. In this context, a methodological approach based on hyperspectral spectroscopy and X-ray diffraction (XRD) method is proposed for the phosphate surface abundance prediction and exploration in a specific geological context in Tunisia. In this study, partial least square regression (PLSR) method was conducted on hyperspectral visible-near infrared (VNIR) and short-wave infrared (SWIR) field reflectance spectra of the collected samples and XRD analysis results for phosphate content prediction. Results revealed that carbonate absorption features in SWIR region could be considered for an accurate estimation of phosphate contents. The generated model has shown an interesting performance with an R^2 of 0.64, an RMSE of 5.52, and an RPIQ of 2.15, using the training samples set. Moreover, X-ray diffraction (XRD) analysis results were used for the validation purposes. The using validation samples set revealed an R^2 of 0.42, an RMSE of 10.29, and an RPIQ of 1.74. All performance coefficients have shown that the generated model can be applied successfully for the content prediction of phosphates. The present study revealed, thus, the contribution of the proposed methodological approach for phosphate exploration in the Chaketma mine site in the Centre West of Tunisia, which can be improved in the future.

Keywords: optical remote sensing, PLSR, XRD, phosphate mineralization, Fluorapatite, content prediction

1. Introduction

Phosphate is essential to all forms of life, which 90% of its consumption is in agriculture. There is no substitute for phosphate, and it is essential to improve crop yields. Otherwise, phosphate rock is considered as a potential new rare earth element (REE) resource according to their world commercial production that is estimated to be 250 million tons per year [1]. Specifically, apatite is included as one of the known host minerals of rare earth elements by exploration geologists and can provide a feasible resource of REE in future, which demand increases nearly 8% per year [2]. Actually, they are vital to green and emerging technologies [3, 4].

Tunisia is well known for its phosphate reserves, since 1887. Gafsa mining basin is the main region of sedimentary deposits exploited in the south [4]. Indeed, Eocene phosphate basins can be classified into three large sites: the Gafsa basin (Metlaoui, M'Dhilla, Moulares, and Redeyef), the Meknassi basin, and the northern basin (Chaketma and Sra Ouertane), which were recently explored as carbonate phosphates. The extracted raw phosphate was processed in Metlaoui, Moulares, Redeyef, Kef Eddour, and M'Dhilla production centers. Particularly, Chaketma phosphate mine presents a potential large-scale, world-class phosphate development asset in Tunisia. The bulk of the phosphate is located at the base of a massive limestone unit close to the top of a high segmented plateau, which is affected by a series of normal faults.

A first study was conducted in a Tunisian semiarid environment for phosphate mapping using both hyperspectral and multispectral remote sensing technologies as an interesting alternative to standard methods, which are routinely applied in mineralogical and geochemical studies [5]. Indeed, remote sensing tools have been well used to map rocks and minerals because of their detailed information, which can be provided periodically. Hyperspectral remote sensing tools at different scale provide higher spectral resolution data that can be used for mineral exploration and mapping [6–11]. Several studies detected minerals basing on their spectral absorption features identified from hyperspectral reflectance data, taken by hyperspectral radiometer. Carbonates, clays, were thus studied and analyzed. Recently, near-infrared (NIR) data were used successfully for the prediction of rare earth elements in the largest uranium-phosphate deposit in Brazil. Three partial least squares regressions (PLSRs): full-spectrum partial least squares (PLS), interval partial least squares (iPLS), and successive projections algorithms for interval selection in partial least squares (iSPA-PLS) were conducted to calibrate the measured spectra to predict rare earth elements in topsoil [12]. PLSR is a statistical and a recent technique that generalizes and combines features from principal component analysis and multiple regression. It is particularly useful in the case of predicting a set of dependent variables from a large set of independent variables, called predictors [13].

Studies that pointing out the prediction of fluorapatite phosphate minerals using VNIR-short-wave infrared (SWIR) spectroscopy data and carrying out PLSR methods are missed. The present chapter shows the preliminaries results, derived from a PLSR-based study that was conducted on a particular semiarid Tunisian context aiming to predict fluorapatite phosphate mineral contents using field hyperspectral spectroscopy and X-ray diffraction (XRD) mineralogical analysis.

2. Chaketma mine site

The Chaketma mine, which is located in the Central West of Tunisia, at 200 km far from the southwest of Tunis, is characterized by its sedimentary phosphates. This facies, which is the most common, comes from three main petrographic constituents: granules (pseudo-oolites), nodules, and organic debris (coprolites).

Consisting of a succession of fractured plateaus, the phosphate deposits of the Chaketma site are bounded to the west by the Ghoualguia anticline by a major NNO-SSE normal fault (N160) (**Figure 1**) controlling the Rouhia hollow. Particularly, in the prospect of Gassaa Kebira, a massive bar of dolomitic limestone, recording a series of Ypresian age and decline of the formation of El Gueria, outcrops thicker than at Gassaa Sghira, in the north of the district, and Kef Oum Ezzine, in the south [14].

Kef El Louz, Sidi Ali Ben Oum Ezzine, Gassaa Kebira, Gassat Ezarbat, Kef El Agueb, and Douar Ouled Hamouda are the six individual phosphate prospects,

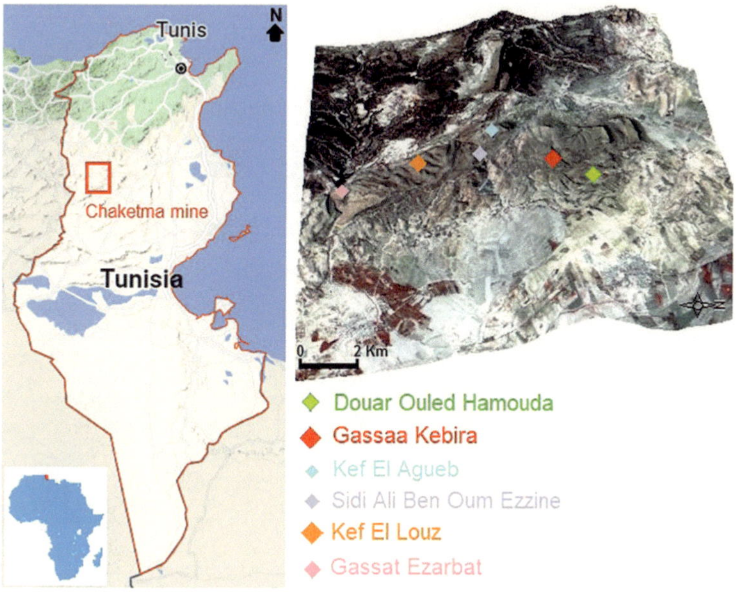

Figure 1.
Localisation of the Chaketma mining site as well as the six phosphate perspectives: Map of Tunisia (left) and combination of Aster red-green-blue (RGB) image draped over the digital elevation model (DEM) of the mine site (right).

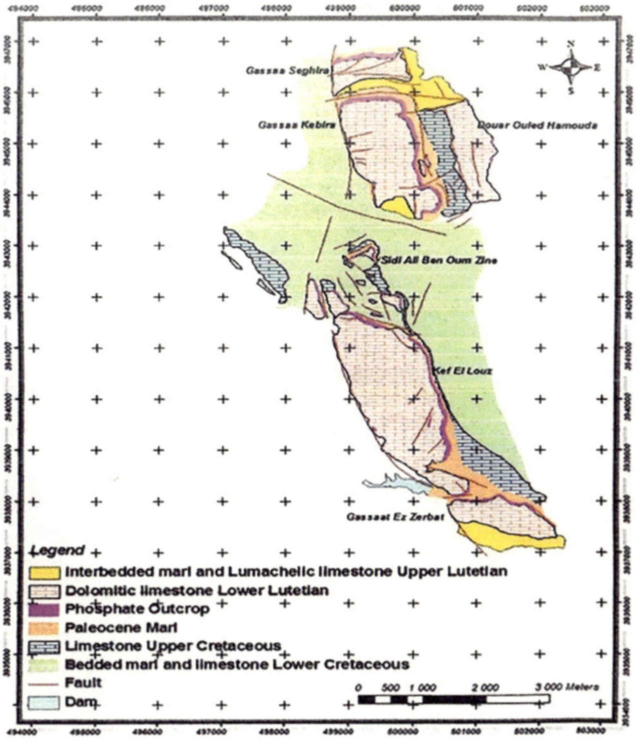

Figure 2.
Geological map of the Chaketma mining site (TMS Company, 2012).

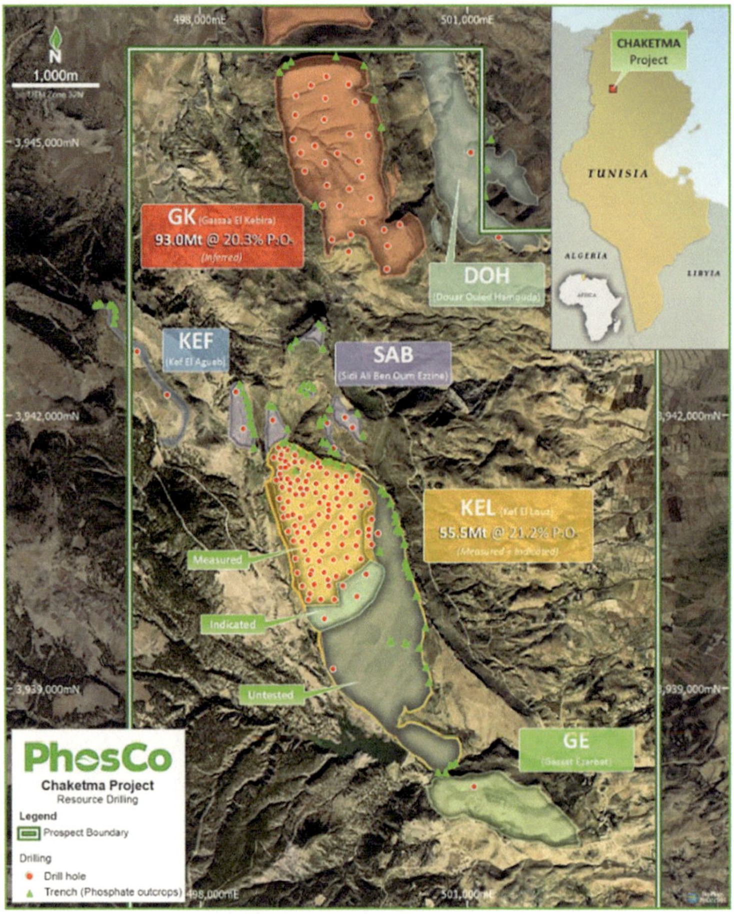

Figure 3.
The six phosphate perspectives (PhosCo Company, 2020).

showing thick mineralized zones with more than 15 m at depth [15], which were determined in 2012 by the Celamin Holdings Company. Tunis Mining Service TMS company estimated, however, the mineralization in the six prospects of the Chaketma mine at about 176 Mt. at 19.5% of pentoxide phosphorus P_2O_5 (**Figure 2**) and 79 Mt. of phosphate at 30% P_2O_5 content. Recently, PhosCo Company has revealed that the deposit contains a resource of 148Mt at 20.6% P_2O_5, confirmed from drilling at only two of the project's six prospects, precisely, 93 Mt. at 20.3% P_2O_5 in Gassaa Kebira and 55.5 Mt. at 21.2% P_2O_5 in Kef El Louz (**Figure 3**). According to PhosCo, the phosphate resource at Kef El Louz is large, shallow and features simple geology. Drilling results have produced consistent, wide, high-grade phosphate mineralization close to surface, with 50% of the prospect's known surface mineralization yet to be drilled [16].

3. Methodological approach

The methodological approach (**Figure 4**) aiming to fluorapatite content prediction is based on both mineralogical and spectral analysis data. Indeed, the gathered phosphate samples from the top surface, after draying and quartering, were the subject

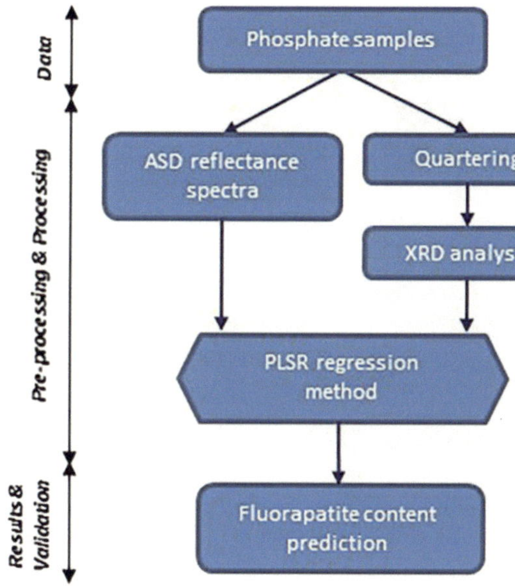

Figure 4.
Flowchart detailing the methodological approach.

of X-ray diffraction (XRD) analysis, using a PANalytical X'Pert PRO X-ray diffractometer, to identify and estimate the abundance of each mineral. Furthermore, all 25 surface samples in different point locations were measured under natural light with the ASD FieldSpec HiRes spectroradiometer (Analytical Spectral Devices, Boulder, Co.). This spectroradiometer, which is fitted with 10°field-of-view fiber optics, operated in the 350 nm to 2500 nm spectral regions with sampling intervals of 1 nm. The mean of the five surface VNIR-SWIR reflectance spectra scans of each sample was preprocessed before using for the partial least square linear regression (PLSR). The prediction content results were validated basing on their comparison with the measured ones by XRD.

4. Results and discussions

4.1 Phosphate mineralogical composition

The interpretation of X-ray diffractograms of all selected samples revealed a high concentration of both calcite and dolomite minerals [5]. The carbonated geological context of the region can explain the presence of these high carbonate mineral abundances. Particularly, dolomite ($CaMg(CO_3)_2$) and calcite ($CaCO_3$) showed high concentrations of up to 97% and reached 68%, respectively. The phosphate mineral, fluorapatite ($Ca_5(PO_4)_3F$, reached 27.21% in the sector of Sidi Ali Ben Oum Ezzine and 34.05% in the sector of Gassaa Kbira. An overview on result statistics of the fluorapatite is showed in **Table 1**. Quartz (SiO_2) was also detected. It showed average to low abundance that can reach 30%.

4.2 Phosphate spectral behavior

According to the identified specific vibrational absorption features on the measured reflectance spectra, minerals were detected in convergence with the

Parameter	Value
Obs	25
Min	0
Max	34.05
Mean	10.79
Median	6.01
Std Dev	10.41
Skewness	0.64
Kurtosis	2.26

Table 1.
Fluorapatite result statistics.

mineralogical results. Carbonate minerals were highlighted thanks to a specific vibrational absorption feature in the short-wave infrared (SWIR) region at 2336 nm due to CO_3^{2-} ion. Dolomite presents, in particular, a displaced absorption feature at 2326 nm instead of 2336 nm, compared to calcite (calcium carbonate), which presents an absorption minimum at 2270 nm instead of 2298 nm (**Figure 5**).

Fluorapatite is, however, featureless within the spectral range from 2000 to 2500 nm [17]. Some spectral features could, however, be considered for a judicious diagnostic and an efficient characterization of phosphate-enriched rock. Indeed, the flatness of rock phosphate spectra as compared to the spectrum of dolomite around the spectral range from 2230 to 2400 nm as well as the absorption kink at 2209 nm have been used for this purpose. The defined absorption features were confirmed basing on the spectrum of a fluorapatite sample that was collected from a Tunisian carbonate geological context. Considering the both carbonate minerals in this case region, the phosphate-enriched dolomitic rock showed a decrease in the value of the absorption depth at 2326 nm as well as for the absorption minimum at 2270 nm. The phosphate-enriched limestone rock also showed a similar behavior but at different wavelengths, at 2336 nm and 2298 nm, respectively. The presence of fluorapatite was still detected according to 2209 nm

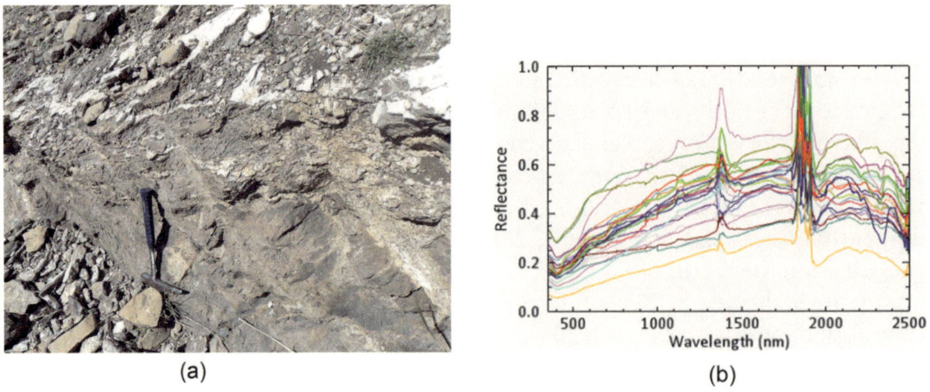

Figure 5.
Phosphate samples: (a) photo of the phosphate surface and (b) ASD VNIR-SWIR field reflectance spectra, taken on the surface of the different sampling points.

absorption feature [5] within both phosphate-enriched dolomitic rock and phosphate-enriched limestone rock spectra.

4.3 Fluorapatite content prediction

4.3.1 PLSR regression method

The PLSR regression method was conducted using the resulting XRD analysis data and the measured VNIR-SWIR hyperspectral data in an attempt to predict fluorapatite contents within phosphate samples. No transformations have been applied on the used reflectance spectra.

Aiming to verify the prediction capability of each generated PLSR model for the training dataset, an optimum number of latent variables (LVs) were determined through the leave-one-out cross-validation procedure [13, 18], which was repeated for all samples to predict fluorapatite contents. The resulting optimal PLSR model was then applied to the validation dataset, which represents 30% of selected samples. Seventy percent of samples were, however, selected for training purpose (**Figure 6**).

According to the previous studies [19], fluorapatite concentrations were sorted in increasing order, interpreted as the optimal method. The performance of the generated PLSR model was assessed through the comparison of predicted content values of fluorapatite dedicated for validation with the measured ones. Thus, the coefficient of determination R^2 in the training and the validation sets, the root mean square errors (RMSE) and the ratio of the performance to interquartile (RPIQ) were calculated. The most significant wavelengths n in the PLSR were justified [20] and used for PLSR model generation.

4.3.2 PLSR linear regression

The generated PLSR model for the fluorapatite content prediction was generated using hyperspectral reflectance spectra and its respective concentrations, estimated

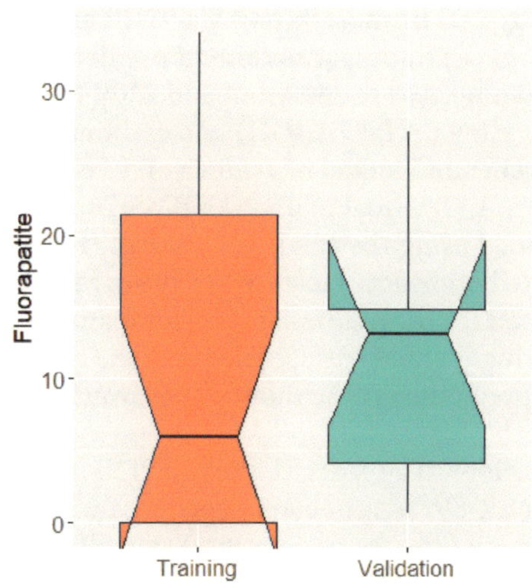

Figure 6.
The selected samples for training and validation purpose.

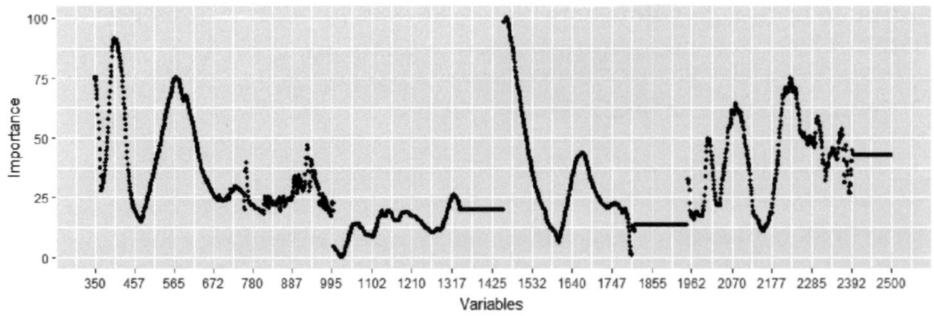

Figure 7.
The most spectral regions used in the PLSR regression for the prediction of fluorapatite concentrations.

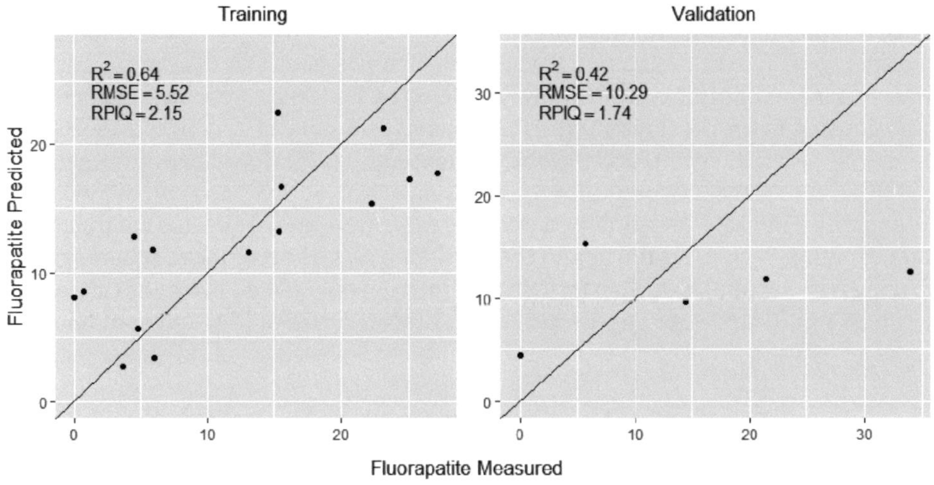

Figure 8.
Accuracies of the PLSR model, generated for the fluorapatite content prediction.

by X-ray diffraction analysis. The most important VNIR and SWIR spectral regions were used by the PLSR for the fluorapatite content prediction of 400 to 450 nm, 500 to 570 nm, 2000 to 2100 nm, 2180 to 2200 nm, and 2300 to 2350 nm (**Figure 7**). The spectral bands corresponding to clay (Al-OH) and carbonate (CO_3) occurrences were particularly used. The generated model has shown an R^2 equal to 0.64, an RMSE of 5.52, and an RPIQ at 2.15. An R^2 equal to 0.42, an RMSE of 10.29, and an RPIQ at 1.74 were, however, determined using the validation dataset (**Figure 8**). It has been shown that these interesting prediction accuracies, which were revealed using these preliminary results, are promoted for the prediction of fluorapatite contents.

The present chapter highlighted the usefulness of the hyperspectral reflectance spectra for the content prediction of the fluorapatite around the Chaketma phosphate mining site.

The detailed methodological approach based on PLSR regression method, which was conducted using VNIR-SWIR reflectance spectra, taken on the surface of the different point sampling, and the corresponding X-ray diffraction (XRD) results, was presented. Results have shown that fluorapatite, the main phosphate mineral in the Chaketma mine, was predicted according to the interesting prediction accuracies, showing an R^2 = 0.64, an RMSE = 5.52, and an RPIQ = 2.15.

A thorough study will be conducted in the future with the aim of further improving the prediction performances of the fluorapatite content modeling, through the application of the data preprocessing methods. Moreover, multispectral image data, such as SENTINEL-2 and Aster data, as well as hyperspectral image data, such as Hyperion, will be used for the mapping of this phosphate mineral. Such results will be of great interest for locating and estimating the phosphate reserve.

Author details

Nouha Mezned[1,2]

1 Faculty of Sciences of Tunis, Laboratory of Mineral Resources and Environment, University of Tunis ElManar, Tunisia

2 Faculty of Sciences of Bizerte, University of Carthage, Tunisia

*Address all correspondence to: mezned_nouha@yahoo.fr; nouha.mezned@fst.utm.tn

References

[1] Shengxi W, Liangshi W, Longsheng Z, Patrick Z, El-Shall H, Brij M, et al. Recovery of rare earth elements from phosphate rock by hydrometallurgical processes – A critical review. Chemical Engineering Journal. 2018;**335**:774-800. DOI: 10.1016/j.cej.2017.10.143

[2] Sinding-Larsen S, Wellmer F. Non-renewable resource issues: Geoscientific and societal challenges. In: International Year of Planet Earth Series. Dordrecht, The Netherlands: Springer; 2012. DOI: 10.1007/978-90-481-8679-2

[3] Owens C, Nash G, Hadler K, Fitzpatrick R, Anderson C, Wall F. Apatite enrichment by rare earth elements: A review of the effects of surface properties. Advances in Colloid and Interface Science. 2019;**265**:14-28. DOI: 10.1016/j.cis.2019.01.004

[4] Sassi S. La sédimentation phosphatée au Paléocène dans le Sud et le Centre Ouest de la Tunisie. Orsay, Paris, France: Doctorat d'Etat es-Sciences; 1974

[5] Mezned N, Fatnassi A, Abdeljaouad S. Potential of ASTER's VNIR, SWIR, and TIR bands for phosphate exploration in central West Tunisia. Journal of Applied Remote Sensing. 2020;**14**(1):014509. DOI: 10.1117/1.JRS.14.014509

[6] Clark RN, Roush TL. Reflectance spectroscopy: Quantitative analysis techniques for remote sensing applications. Journal of Geophysical Research. 1984;**89**:6329-6340

[7] Gaffey J. Spectral reflectance of carbonate minerals in the visible and near infrared (0.35-2.55 um): Anhydrous carbonate minerals. Journal of Geophysical Research. 1987;**92**:1429-1440

[8] Clark RN. Spectroscopy of rock and minerals. In: Rencz A, editor. Principles of Spectroscopy, in Manual of Remote Sensing. New York: John Wiley and Sons Inc.; 1999. pp. 3-58

[9] Pazand K, Sarvestani J, Ravasan R. Hydrothermal alteration mapping using ASTER data for reconnaissance porphyry copper mineralization in the Ahar area, NW Iran. Journal of the Indian Society of Remote Sensing. 2013;**41**:379-389

[10] Kruse FA. Integrated visible and near infrared, shortwave infrared, and longwave infrared (VNIR-SWIR-LWIR), full-range hyperspectral data analysis for geologic mapping. Journal of Applied Remote Sensing. 2015;**9**(1):096005

[11] Gomez C. Sensitivity of clay content prediction to spectral configuration of VNIR/SWIR imaging data, from multispectral to hyperspectral scenarios. Remote Sensing of Environment. 2018;**204**:18-30

[12] Maia AJ, Bezerra da silva Y, Araujo do Nascimento C, Veras G, Escobar M, Cunha C, et al. Near-infrared spectroscopy for the prediction of rare earth elements in soils from the largest uranium-phosphate deposit in Brazil using PLS, iPLS, and iSPA-PLS models. Environmental Monitoring and Assessment. 2020;**192**(11):1-14. DOI: 10.1007/s10661-020-08642-2

[13] Wold S, Sjöström M, Eriksson L. PLS-regression: A basic tool of Chemometrics. Chemometrics and Intelligent Laboratory Systems. 2001;**58**:109-130. DOI: 10.1016/S0169-7439(01)00155-1

[14] Briki H. Structural evolution and tectonic style of the Tunisian

central atlas; role of inherited faults in compressive tectonics (Ghoualguia anticline). Tectonophysics. 2018;**731-732**: 48-63

[15] Celamin Holdings NL. Annual Report. 2013. ABN 82 139 255 771. p. 71

[16] PhosCo. 2020, www.phosco.com.au

[17] Lane MD, Dyar MD, Bishop JL. Spectra of phosphate minerals as obtained by visible-near infrared reflectance, thermal infrared emission, and Mössbauer laboratory analyses. In: 38th Lunar and Planetary Sci. Conf. League City, Texas: LPI; 2007, Contribution No. 1338. p. 2210

[18] Viscarra Rossel RA, Walvoort DJJ, McBrateney AB, Janik LJ, Skjemstad JO. Visible, near infrared, mid infrared or combined diffuse reflectance spectroscopy for simultaneous assessment of various soil properties. Geoderma. 2006;**131**:59-75. DOI: 10.1016/j.geoderma.2005.03.007

[19] Dkhala B, Mezned N, Gomez C, Abdeljaouad S. Hyperspectral field spectroscopy and SENTINEL-2 multispectral data for minerals with high pollution potential content estimation and mapping. Science of the Total Environment. 2020;**740**:18 p. art. 140160, ISSN 0048- 9697. DOI: 10.1016/j.scitotenv.2020.140160

[20] Chong IG, Jun CH. Performance of some variable selection methods when multicollinearity is present. Chemometrics and Intelligent Laboratory Systems. 2005;**78**:103-112. DOI: 10.1016/j.chemolab.2004.12.011